SO LONELY

HILDE ØSTBY

Translated by MATT BAGGULEY

So Lonely

Our Desire for Community—
And What Drives Us Apart

GREYSTONE BOOKS
Vancouver/Berkeley/London

First published in English by Greystone Books in 2025
Originally published in Norwegian as *Kart over ensomheten* [Map of Loneliness],
copyright © 2022 by Cappelen Damm
English translation copyright © 2025 by Matt Bagguley
"Moss" by Kjersti Bjørkmo is reprinted with permission of the author.

25 26 27 28 29   5 4 3 2 1

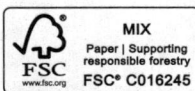

Greystone Books Ltd.
greystonebooks.com

Cataloguing data available from Library and Archives Canada
ISBN 978-1-77840-002-5 (cloth)
ISBN 978-1-77840-003-2 (epub)

Editing by Brian Lynch
Proofreading by Crissy Boylan
Jacket design by Javana Boothe
Jacket photograph by Palana997/istock.com
Text design by Fiona Siu
Printed and bound in Canada on FSC® certified paper at Friesens.
The FSC® label means that materials used for the product have
been responsibly sourced.

Greystone Books thanks the Canada Council for the Arts, the British
Columbia Arts Council, the Province of British Columbia through
the Book Publishing Tax Credit, and the Government of Canada
for supporting our publishing activities.

This translation has been published with the financial support of NORLA.

Canada | N NORLA

BRITISH COLUMBIA   BRITISH COLUMBIA
                    ARTS COUNCIL
                    An agency of the Province of British Columbia

Canada Council   Conseil des arts
for the Arts   du Canada

MIX
Paper | Supporting
responsible forestry
FSC
www.fsc.org   FSC® C016245

Greystone Books gratefully acknowledges the xʷməθkʷəy̓əm (Musqueam),
Skw̱x̱wú7mesh (Squamish), and səlilwətaɬ (Tsleil-Waututh) peoples on
whose land our Vancouver head office is located.

Contents

Why loneliness is the hunger of the soul, and why we cannot bear to be abandoned

THIS TALE ABOUT loneliness began when I was reading a bedtime story to my six-year-old daughter on a dark autumn night during the pandemic. With only a small bedside lamp illuminating the pages, we huddled over an illustrated book by Tove Jansson about lonely Toffle, who is frightened at night by the howling of the Groke. My daughter listened enthralled and delighted by the colorful pictures. But when we finished the story and closed the book, I felt strangely uncomfortable. Something wasn't right. Why had this children's book been scary to me—but not her?

I must keep going, I thought, toward the powerful sense of being lost the book so clearly described. It's like a compass point! There was something about the story of Toffle that I didn't understand and wanted to investigate: Why was he so lonely? Who was the Groke, and were there more of them? I knew that *Who Will Comfort Toffle?* is "just a children's book," but I also knew that there's a dark truth hidden in the brightly colored pictures; loneliness is a real and modern problem, not a children's fairy tale. So I decided that I would take a closer look.

A few days later, as my family and I were eating baked salmon for dinner, I mentioned it to my husband. He sat there nonchalantly balancing a piece of salmon on his fork—while our daughter picked at the food with the usual look of disgust on her face—and said, "Loneliness? Yes, why not? I mean, you're probably the loneliest person I know," in a very direct and matter-of-fact way, not snarky or mean. It was said as though it should have been obvious to me. But it wasn't.

Since then, I've been trying to figure out what he meant. I have plenty of friends who invite me to dinner and birthday parties, thousands of "friends" on Facebook, I live in the middle of a bustling capital where I hang out with my neighbors and help at the school jumble sale. No one has ever called me lonely before. But I like *being* alone, and neuroscience shows that walking alone and daydreaming every day can be highly beneficial. Sometimes we all need to be like Jansson's wandering nomad Snufkin: *"It's the right evening for a tune*, Snufkin thought. A new tune, one part expectation, two parts spring sadness, and for the rest, just the great delight of walking alone and liking it." Walking alone, immersed in our own thoughts, can positively affect our memory, make us more creative, and help us get to know ourselves better. We humans live in highly complex social structures and need to understand our place in the world, who we are, and how important others are to us. So being *alone* is too imprecise a definition for loneliness, and perhaps doesn't mean *lonely* at all. Sometimes we just *must* be alone; being alone in your own head is one of life's necessities—a good one: I love walking and thinking in absolute silence and have written a whole book about daydreaming and mind-wandering. But my husband knows all this, and that wasn't what he meant—that I often disappear into my own thoughts. It's what I do for a living.

Nor do I think he was talking about existential loneliness. This most fundamental sense of loneliness applies to us all; it is loneliness's magnetic north pole, something every human being in the world is equipped with: you realize, with a sudden shudder, that you will never fully understand other people and that they will never fully understand you, that you will die alone, and that your life is of little importance in the grand scheme of things; the stars dancing across the night sky are a reminder that you are a minute speck of dust in a colossal universe. These thoughts arise when you are a teenager, they are material for poetry and literature, the kinds of questions that sit at the very core of the world's great religions; they reflect the shock and absurdity of being alive on this planet with an awareness of our own mortality.

Loneliness is often defined as a feeling of "not getting the social contact we desire," but what that actually means is more unclear. Because what is it we really want from human contact, what is it that fulfills our longing for other people? It also turns out that loneliness can be linked to very specific situations— such as when you change schools, get a new job, move to a new place or into an institution; you are vulnerable to loneliness in all of life's transitions. But it doesn't have to be like that. And it's not like that for me. I like transitions.

The question I would like to answer is: What causes *dysfunctional* loneliness—the fissures between us, all the things that keep us apart for no reason, the loneliness that suddenly makes Toffle barricade himself behind his door when the Groke starts howling in the night, that makes him jump at the sound of the hemulen's footsteps? This form of loneliness is a virtual mystery to science, just as it is to me. Yet it occurs in all walks of life and in all types of people. Before the Covid-19 pandemic, it was known that, at any given time, 20 percent of Norway's population was lonely, and what's disturbing is that over the last few decades loneliness has steadily increased among young people. During the lockdowns, the number of lonely teenagers in Norway increased at an alarming rate, and now, in the aftermath of all this isolation, the number is as high as 70 percent among people between sixteen and nineteen. But even before this, numerous studies showed that the rates of loneliness were frighteningly high: almost 50 percent of adults in Great Britain either always or very often feel lonely. The BBC found that half of Britons over sixty-five say a pet or the TV is their best friend. The British government calculated that loneliness costs society 3.5 billion pounds a year, which led its then prime minister Theresa May to appoint Britain's first minister for loneliness, Tracey Crouch, in January 2018, without solving the problem. In the US, a recent survey conducted by Cigna shows that almost 60 percent of Americans think that no one really knows them well, a figure that is higher among the Hispanic and African-American populations. Denmark, Australia, and Great Britain launched large campaigns, but despite rising loneliness, these were temporarily shelved during the pandemic.

We also feel alienated and disconnected in the very place we spend most of our lives: globally, almost 80 percent of employees say that they do not feel a sense of belonging to their workplace, 60 percent are not engaged at work, and 19 percent are directly unhappy. In the US, 40 percent of all employees do not have a single friend at work—and this figure is expected to rise now that we are increasingly working from home. Participation in associations, organizations, and leisure activities is also in decline, something Robert Putnam examined in his book *Bowling Alone*, which claims that since the 1970s membership in organized activities in the US has fallen between 25 and 50 percent. Putnam believes that the erosion of this social glue is a threat to democracy.

Loneliness appears in our cities, where we live closer together than ever before. Soon, up to 70 percent of the world's population will be living in cities, and the number of people living alone is increasing in cities across the world, somewhere up to 50 percent, as is the case in Oslo, my hometown—a figure that makes you wonder if cities are lonely by design, because where we find high populations, we also find loneliness. This is a mysterious form of loneliness, given that having access to lots of people should alleviate it, not make it worse. In her book *The Lonely City: Adventures in the Art of Being Alone*, Olivia Laing writes: "You can be lonely anywhere, but there is a particular flavor to the loneliness that comes from living in a city, surrounded by millions of people." Loneliness increases where people have little trust in each other, such as in the US. But that doesn't explain everything, because loneliness is found where people are known to be very trusting of each other and the surrounding community, like Japan for example; yes, actually *more* than other places, which somewhat undermines that theory too. Perhaps lonely city folk are just like Toffle. He doesn't trust his surroundings and is frightened simply by the noises outside his own house. So why is it that we can suddenly become afraid of each other? And who *will* comfort Toffle? And why do some people go through life rarely feeling lonely, while others are totally consumed by it?

I think it was this mysterious loneliness that my husband had in mind and had so straightforwardly referred to at the kitchen table: This thing gnawing at me. This thing that's so difficult to grasp. It's something that shouldn't be there, because although I am surrounded by people, I'm still afraid somehow. Afraid of losing perhaps, of being judged, of not being good enough, afraid of not being liked or accepted as I am—but this feeling is just as hard to articulate as the vague discomfort I experienced when I read the book about Toffle. So where does this feeling come from?

I suddenly realized that to grasp the essence of loneliness, I needed to understand a lot more about the things that connect us to each other. I needed to find out what it means to have a real connection with other people—how we cling to each other, how we wrap ourselves around each other like warming quilts and blankets. Because this also says something about how things can turn so cold between us. I had to get to the bottom of it, not just to understand myself, but to understand what happened when a pandemic threw the entire world into involuntary isolation. It's as though, since early 2020, we've all—all seven billion of us—been unwilling participants in a colossal loneliness experiment. We've been trapped indoors, often with no way of seeing each other except via the internet, and it has done something to us, as though meeting via Zoom or hearing a voice on the phone isn't enough. We need more!

The dire situations caused by the pandemic have created fertile ground for loneliness: For those suffering in the world's most poverty-stricken areas, without access to medical care. The vulnerable children who found themselves trapped in the same home as the adult who is mistreating them, with no chance of being seen or helped by anyone beyond those four walls. The many students who have developed psychological problems because online classes have been the only thing to break the monotony of staying in their tiny bedsits. The employees who have gone months without seeing their colleagues. The elderly who have been prevented from cuddling their grandchildren, unable to feel a soft, tender cheek pressed hard and eagerly against their own. And the many people who have been forced to say final farewells to dying relatives via FaceTime. All this, and much more. Today, on our tiny cellphone screens, we can see almost everything in this world and call virtually anyone living in it. But we still have loneliness. It lives hidden in our midst.

But perhaps calling all these lonelinesses the same thing only creates an illusion of similarity. The loneliness of a privileged woman, such as me, living in a safe Scandinavian city is of course not the loneliness experienced by a traumatized child trapped in a refugee camp. A woman suffering from dementia in a nursing home experiences a very different loneliness from that of a young man who has been seduced by right-wing-extremist websites, doesn't she? And just to complicate things further: although something might appear to be loneliness, none of these situations necessarily *make* us lonely. Even after the pandemic, some people emerged from these years of lockdown scarred and wounded, while others left this laboratory of solitude and brushed off their isolation as if it were just an unfortunate incident. Loneliness is an individual and personal experience. We know that people are lonely because that's what they answer when asked by researchers and statisticians. It's not a diagnosis, and it cannot be seen from the outside.

So what is loneliness, essentially? I *do* have a few leads to follow. Loneliness originally had a function and is strongly connected to what makes us human. It is actually quite a logical feeling. Like everything we humans are equipped with in terms of needs and emotions, a painful feeling like loneliness is also something we have developed: feeling lonely improved our chances of survival when we were living on the savanna and the human brain was evolving into what it is today. Loneliness is the compulsion that arises from being afraid of losing your community. It is linked to our fear of being ostracized from the group. When the body secretes large amounts of stress hormones due to us feeling lonely, it does so because we must immediately try to return to the community. Back on the savanna, being abandoned was synonymous with death.

"Physical pain protects the individual from physical dangers," writes John T. Cacioppo. "Social pain, also known as loneliness, evolved for a similar reason," he continues. Cacioppo researched loneliness at the University of Chicago for twenty-three years and created the field of social neuroscience, which shows how we are totally dependent on each other. We cannot talk about the human brain in isolation; it is something that works best when we are with others. His research shows that the brain is, above all else, a social organ.

Through his extensive and groundbreaking research, Cacioppo has documented what prolonged loneliness does to us. Short-term loneliness serves as a reminder that we need to return to the community. Just as hunger makes you eat—since without food you would die—short-term loneliness reminds you that you need people. That is why the consequences of chronic loneliness and the stress it causes are so harmful: persistent loneliness leads to overeating, inadequate sleep, drug and alcohol use disorders, attachment issues, stress, autoimmune diseases, and impaired concentration.

Cacioppo believes that the opposite feeling to loneliness is depression, and that the two feelings are actually meant to regulate each other: loneliness makes us actively seek out other people, whereas depression makes us withdraw and reevaluate a situation before embarking on a new advance into the social jungle. But originally loneliness and depression counteracted each other quickly. They are not feelings we can handle for very long.

"The limited cognitive powers of the earliest hunter-gatherer societies, and the harshness of their environment, would not have allowed them the luxury of long bouts of passive melancholy, ambivalence and soul-searching," writes Cacioppo. "Over many millennia, however, with increasing intellectual and psychosocial complexity, a simple sequence of 'go/stop/go again' has evolved into a vicious cycle of ambivalence, isolation and paralysis by analysis—the standoff in which loneliness and depressive feelings lock into a negative feedback loop, each intensifying the effects and persistence of the other."

When we lived in small tribal communities, and a member of the group strayed too far while searching for new habitats or hunting grounds, the feeling of loneliness would soon compel him to return. And if the social interacting went badly for this early *Homo sapiens* and his attempts to engage with others were failing, he would retreat, feel melancholy, reassess the situation, and try again. This can be a well-functioning interplay, between the longing for others that makes us seek company and the corrective, short-term feeling of depression that makes us adjust back to being as normal as possible.

Paying attention to social cues and trying to behave "like the rest of the pack" is a pure survival strategy, considering that exclusion would be fatal; yes, you could end up as lion food if your pack didn't want you around. Being accepted as normal has been an essential part of staying alive. From as early as five or six years of age, we start internalizing the rules and cultural codes we perceive, from the world around us, about what's normal and what isn't. Even my seemingly independent six-year-old has started noticing what "everyone else" is doing: On Halloween, she went to school dressed as a slightly ragged-looking lion—it was her favorite costume and had worked just fine in kindergarten. But when we entered the schoolyard as the bell was about to ring, we saw that nearly all the other first-grade girls were dressed as witches. An anguished look spread across my daughter's face. "Okay, but look, there's a cat!" I said, trying to spin the situation positively. "And a tiger too! And a princess! There are lots of different costumes!" But most of the girls were witches, and that was all my daughter could see. "I should have been a witch," she insisted. And when she returned from school one day just before Christmas with a wish list addressed to Santa Claus in her bag, there was only one thing written on it: "I want to be normel." As painful as it was to read—because I value all my daughter's strange quirks and foibles—I know that this desire to be "normel" is a manifestation of an essential survival skill.

Twin studies have shown that approximately 50 percent of our traits and characteristics are those we are born with. This means that our capacity for loneliness is something we are partly encoded with, but it can be equally shaped by the environment we grow up in. Cacioppo believes that we all experience loneliness in different ways because a human group would have had better conditions for survival if it consisted of several types of people: Someone had to maintain the core of the community and the close social ties that are so vital to us. At the same time, humankind has always needed adventurers, people who would expand the areas of unfoldment, who travel to unknown places or share bold new visions and discoveries. We have of course needed both explorers and those who build social institutions.

It is the very secret of humanity's success, why we are the most successful species on the planet—because we are so diverse. This is one of the positive effects of feeling lonely: loneliness structures groups of people so that some can look after the core of the community while others go looking for new opportunities. My great-grandfather, who was a whale hunter and traveled into the unknown in the Antarctic Ocean, was no doubt able to endure the lack of constant social confirmation and connection far longer than any of my other relatives, who devoted themselves to farming or were housewives. And yes, the latter of these categories applied to most of my ancestors, as it did to the rest of society. Cities are not held together by adventurers and innovators, despite how much our culture glorifies them. At the heart of a society are those who look after the community: nurses and teachers and politicians and theater managers, bus drivers and musicians, cleaners and kindergarten staff. Very few of us head for unknown shores.

Chronic loneliness, however, does not seem to have any such function. Professor Cacioppo believes that it is a result of the civilization we have constructed: modern life allows us to develop chronic loneliness without it really being something we can handle. Chronic loneliness sets off a negative spiral that creates depression and anxiety, low self-esteem, social anxiety, pessimism, and a fear of being negatively judged, all of which lead to withdrawal from community. Cacioppo's research shows that we humans become more distrustful of our surroundings if we are lonely for extended periods, develop anxiety and depression, and withdraw from the community we so desperately need. We become so-called hypervigilant and read other people's faces in a nervous and overinterpreting way. We misinterpret; we become restless and afraid. People who are chronically lonely have no typical characteristics besides the fact that prolonged loneliness leads to them pushing others away, instead of connecting with them. In that way, loneliness creates even more loneliness. Because that is what happens: prolonged loneliness becomes even more difficult to break out of; it becomes self-reinforcing. If you start moving into a spiral of loneliness, you can feel shame and experience insomnia and suffer various addictions, which then reinforce the loneliness and make deep, genuine contact with other people even more difficult to attain. In that respect, loneliness is somewhat paradoxical—it's initially a feeling that's supposed to force us into the community, but if it becomes a persistent feeling, it pushes us further out of the community.

But not only that. In one experiment, participants were grouped in pairs and told in advance that their test partners would either be lonely or have a large social network. Those who had been described as "lonely" were immediately perceived as less interesting by the other person. In other words, not only is loneliness reinforced by the lonely person themselves, but it is also repellant to others, something that compounds the loneliness further still. Backing away from a lonely person seems therefore logical, since that person may already be at risk of being expelled from the group, and we would sooner avoid being caught in the slipstream. And since loneliness is dangerous, it is also fraught with shame. In surveys, women report more loneliness than men and identify with the feeling of loneliness, while men respond affirmatively to having all the characteristics of loneliness: men behave lonely. So perhaps loneliness isn't so unevenly distributed between the sexes; it's simply not as easy for men to talk about, perhaps because it's perceived as a sign of weakness, the mere word *loneliness* drips with shame. Or perhaps, as I do, men find it too vague and difficult a feeling to grasp.

But there is more to loneliness than fear and shame. Research on loneliness has also helped test the almost watertight bulkheads that have been placed between body and soul in today's health care system. Chronic loneliness has a very physical effect on us. In 2010, Julianne Holt-Lunstad, a psychology professor at Brigham Young University, presented a startling discovery—one that would make international headlines and lead to her research paper being cited over five thousand times by other researchers. Holt-Lunstad was actually trying to understand more about the causes of premature death. In her search for environmental factors that are harmful to our health, and those that make us healthy, she looked at 148 studies with a total of over three hundred thousand people. It was an enormous meta-study, pooled together using preexisting studies, to look for larger trends. And she found them. The surprising common denominator for all those who were in good health—when all other factors were considered—was that they had strong social ties.

"Being connected to others socially is widely considered a fundamental human need—crucial to both well-being and survival. Extreme examples show infants in custodial care who lack human contact fail to thrive and often die, and indeed, social isolation or solitary confinement has been used as a form of punishment," Holt-Lunstad says. "Yet an increasing portion of the U.S. population now experiences isolation regularly."

Holt-Lunstad found that loneliness increased the risk of premature death by 26 percent, and the research team was able to calculate that chronic loneliness is worse for your health than smoking fifteen cigarettes a day. This is because loneliness triggers acute stress reactions in the body, a result of the fear of exclusion and death that John Cacioppo described so clearly. Stress is something we humans can only really handle in small amounts, because it wears the body down. It is meant to get us out of a life-threatening situation quickly, so that we can return to being non-stressed. When we experience extreme stress, our immune and digestive systems are put on hold, and we think short-term and not creatively, because we just want to reach safety as quickly as possible. But if the stress becomes chronic, it will cause sleeping problems and low-intensity inflammations that lead to autoimmune diseases, diabetes, heart and vascular disorders, and an increased likelihood of cancer. Holt-Lunstad's research shows that even ambivalent relationships will provoke high blood pressure and anxiety, compared with nice, secure relationships. "Toxic stress" is close to becoming part of the medical terminology in mainstream research and is increasingly being used when documenting what feeling excluded and unloved does to the body. The earlier the exclusion, rejection, and violence begins, the more the body is broken down by the toxic stress that eats away at us from within. The research on loneliness therefore reveals how our emotions and physical health are intrinsically linked. It's not "just in your head." Being lonely really changes your body.

So what actually is loneliness? Why is Toffle so lonely? Am I really like him? And more importantly for me: If I *am* like Toffle, is it something I'll pass on to my daughter? In the novel *One Hundred Years of Solitude*, Gabriel García Márquez tells the story of Aureliano Buendía and his descendants. Buendía, a Colombian rebel colonel whose soul is turned increasingly cold by a civil war, returns to his home village wrapped in a wool blanket, despite it being midsummer. Eventually, he even rejects his own mother: "It was then that he decided that no human being, not even Úrsula, could come closer to him than ten feet. In the center of the chalk circle that his aides would draw wherever he stopped, and which only he could enter, he would decide with brief orders that had no appeal the fate of the world." The story of Aureliano Buendía shows the close relationship between power, terror, violence, and loneliness. Terror is created by loneliness and creates more loneliness. Radicalization and extreme ideologies tear us apart when we could have been a community. But Aureliano also passes his loneliness on to his children and grandchildren, "because races condemned to one hundred years of solitude did not have a second opportunity on earth," writes Márquez. Since I read the book as a teenager, this sentence has burned itself into my memory. It seemed so familiar. Could that be where *my* loneliness comes from? I have often wondered. Is my loneliness inherited?

Just as the pandemic was finally subsiding and we began dropping the masks and safety rules, my father became infected by Covid-19. And then he died. I don't know much about his sense of loneliness because we never talked about it, but I do know that my father had an autoimmune disease that caused the white blood cells in his body to attack his lungs, which in turn made him extra vulnerable to Covid. So when he caught it, his lungs collapsed, and he died on a respirator in the spring of 2022. And when he passed away, all the connection points to my past vanished with him. It was my father who looked after me when I was growing up, and now he was gone. Grief is one of many bleak places on the map of loneliness. When a loved one dies, your moorings to other people come undone. I became socially awkward; the grief made me want to withdraw, to be alone. But it turned into an extra pressing issue, because I felt genuinely seen and loved by my father—and if I already felt this affinity with my father, why was I ridden with such an acute sense of loneliness that my husband thought I was the loneliest person he knew? Had my father been a lonely man and passed his loneliness on to me? I had to go back to the whale hunter, my great-grandfather, and trace the family line until now, to me. This would amount to over a hundred years of loneliness. If loneliness is a dark wave from the past that washes over us on the shore of the present, then I must, and will, be a breakwater. This loneliness will stop with me.

So this book perhaps didn't start when I read the story of Toffle to my daughter, but on a July evening in 1975, when I was born at Oslo University Hospital. Perhaps it began on November 29, 1902, when my great-grandfather was born, because I'm now starting to feel that his life was like a stone dropped in a lake and the ripples have spread and reached all the way to me. In future, when my daughter reads *Who Will Comfort Toffle?*, she will still not know what this children's book is really about: unbearable loneliness and the search for comfort. And she will not have to confront the Groke, who in Moominvalley is loneliness personified. My daughter will always know the way home. And this is my map for her. An unclear map, covered in white spots, because loneliness is so difficult to grasp, so hard to get an overview of. We could almost say it's a map of Moominvalley, where lonesome Toffle goes around plagued by the Groke's howling. Loneliness is a dark landscape—encircled by violence, rape, terror, grief, trauma, suicide, racism, bullying, concealment, rejection, shame, poverty, and lying—with many small paths and stories leading into it. Stories about being forgotten and disregarded, about being overlooked, neglected, and cursed—and about how the Groke became so cold. You will find the 2011 Utøya terrorist attack here. But you will also find a bridge, because there are many glimmers of light. Rather than eroding my faith in humanity, writing this book has boosted it.

The forces striving to end loneliness are far greater than those trying to drive people apart: in the heart of Moominvalley there is a calm, nurturing mother figure called Moominmamma, who offers help, care, and hot meals to passersby, including a little girl called Ninny, who, as a result of being unloved, has become invisible. In the real world, there are far more Moominmammas than there are Grokes. And I know, from my own experience, that an invisible child can become visible. When it comes to people, anything is possible, from the most wonderful to the most dreadful, even murder and terror on an ordinary summer day. So perhaps this book actually started with a bang, one July afternoon in 2011, when I was lying on the couch in my Oslo apartment and heard what sounded like thunder. I had just woken from a nap, and just a few minutes later my telephone rang. It was my friend Ada, who worked in one of the government buildings in the city center.

"I just walked through it," she said. "I just walked through the government quarter!" I could hear from the familiar noise of screeching rails in the background that she was on the tram. And she was clearly upset.

"It was a bomb, did you hear it? I just walked through the building that got bombed. I was there!" said Ada.

She was my friend. She was also an invisible child. This is the story of her and of us others who are lonely—the story of how we will become visible again.

2

Why the eyes really are the mirror of the soul

THERE, HUNCHED OVER a book, reading that night's fairy tale, I encountered the invisible child again in Tove Jansson's tales from Moominvalley. When I last read the story, I hadn't really understood what it was about—I was still a child at the time. But now, when I read it aloud to my daughter, it finally became clear. There's a good reason why Ninny became invisible: she lives with a person who doesn't like her. Not being seen, figuratively speaking, *makes* you invisible, and if you're invisible, you cannot be looked after. So "The Invisible Child" is a story about intense loneliness. It is the story of me.

"No man is an island, entire of it self; every man is a piece of the continent, a part of the main," wrote the poet John Donne in 1624. But when we lose our connections to other people, the fragile bridges we build to this continent collapse. So my human map is of an island—after all, islands have been symbols of loneliness for centuries. And we can begin with the eyes. When all your connections to other people are gone, you become invisible, and sometimes, if you're excluded from a community, this can be done to you in such a subtle way that it's hard to remember why it occurred. It usually starts clearly and physically with a look, before going on to make you feel like nobody can see you, to show that you are not part of the group. And it has become clearer to me now that a single look can be enough to make you feel unwanted. Eye contact is a totally crucial part of how we bond with each other.

Babies who can see from birth look predominantly for faces and eyes. Infants can rapidly detect faces hidden in chaotic photographs, according to a 2019 experiment that was conducted on babies between three and twelve months of age. They first look for faces in all contexts, even faces that are obscured. And that's not so strange, because a baby won't survive long without the protection of an adult who understands their needs. The renowned developmental psychologist Donald Winnicott actually believed that we cannot refer to a baby as a separate entity: "There is no such thing as a baby. If you show me a baby you certainly show me also someone caring for a baby, or at least a pram with someone's eyes and ears glued to it," he said. A baby is never alone; they will die otherwise, so contact with an adult is essential for them to live.

A baby quickly recognizes the faces of their most immediate caregivers, and after only a short time they will prefer the faces of their caregivers to those of other people. These people will be key to the baby's further development—and how they *look* at their baby can be of profoundly decisive significance. We know this because of psychologist Edward Tronick. In 1975, Tronick created an experiment about facial expression that is still used today. The importance of deep eye contact and recognition from a caregiver is documented in his famous "still face" experiment. The setup of the experiment may sound quite simple: a mother is asked to look at her child for two whole minutes, and instead of responding she must maintain a totally expressionless face.

"At the time, people thought babies couldn't participate in social interaction," he later explained.

With his experiment, Tronick was able to show how crucial face-to-face contact, and eye contact in particular, is for very young children. The situation that unfolds during "still face," and which has been repeated in countless experiments since, is heartbreaking. First, the mother and child have a meaningful exchange of sounds and words, of looks and smiles, of pointing and eye contact. They share friendly sounds and gestures. The mother then turns away from the child, before turning back again with a totally "dead" face. She remains like this for two minutes, while the child tries various strategies to bring back their familiar, loving mother, the caregiver who normally responds and communicates with looks and gestures. The baby reaches out, trying to engage their parent emotionally by smiling or pointing. The baby then turns away in frustration, then starts to cry or display other signs of stress until the child finally gives up completely. And all this unfolds in just two minutes of cold eye contact—120 seconds that changes the entire dynamic between them. The mother then returns to how she was before, and the child shows clear signs of relief and joy that everything is back to normal: lively eye contact, facial expressions, wordless communication, and happiness.

This method has become a standard test for assessing how carers and children interact. It has been used to investigate children with Down syndrome, children with hearing difficulties, children of parents with different diagnoses, especially depression—as well as cultural differences, gender differences, autism, attachment styles, and communication differences. The test looks for what kind of expectations the child has of their caregiver in terms of emotional availability.

"Eye contact is important, but we also communicate via several senses," psychologists Ida Brandtzæg and Stig Torsteinson explain to me. "Babies are biologically designed to orient themselves toward a person's voice, speech, and face; their sensory apparatus quickly becomes discriminating between their parents and other people, between the smell of people they do not know and the smell of a primary caregiver."

Brandtzæg and Torsteinson are experts in attachment psychology and have worked on studies of children and with parents wanting to deal with attachment problems. This field of psychology, which has become their specialty, has been significantly shaped by the "still face" test. Eye contact is important not only for attachment, but also as a way of permitting small breaks in the contact. We can't be connected to our children all the time, nor will a child want full attention from an adult all the time. If you stare into your baby's eyes continuously without respecting its need for a break, the child will most likely become stressed, and if done consistently, you will create an insecure relationship. The child will attempt to withdraw, and if this way of relating to the child continues, the withdrawing will become a recurring pattern that follows the child into adulthood. We all need time to ourselves. In a healthy relationship, we constantly break contact a little, we disconnect and connect, repair and change the relationship in a continuous flow of contact and non-contact. What's perhaps most decisive for the child's development is how good we are at reestablishing contact, not the break in contact itself. The attachment psychologists call this microrepairing.

"What we can clearly observe is when the caregiver is unable to resume contact—when the mother or father fails to completely return to the child after the two minutes of having a totally expressionless face. Sometimes it can take a long time before she or he engages with the child again, with its eyes and face. The child may then feel rejected and feel alone as a result," says Brandtzæg.

"If a parent spends too long engrossed by their cellphone, and during periods when the child needs contact, the child will find that it cannot emotionally engage the adults closest to it, which is of course frustrating," she continues. "And that worries me. It makes children alone, and eventually, if they experience it often, they will make do with being lonely. This applies to all types of rejection."

This type of loneliness is really, in the deepest sense, about a child feeling inadequate or unimportant to their parents. If a lack of emotional availability is consistent in the relationship, we're talking about emotional neglect. A neglected child is an unprotected child. It does not matter why there is a lack of involvement with the child; not being seen makes the child feel abandoned, which in turn is associated with increased stress.

"The probability of this is greater if parents have a serious mental illness or are emotionally unavailable for other reasons." Torsteinson explains. "The child will eventually give up on getting this contact. These children do not return as quickly to the carer during a 'still face' test. They give up and turn away from the carer more often. What we see is that the child's expectations of their carer are affected by the availability of the adult."

A child who finds that their parents turn away a lot and don't recognize their need for ongoing communication may become more inattentive and develop problems with regulating emotions and behavior and with trusting and relating to other people. A number of scientific studies show that depressed mothers are unable to engage as strongly in the social game with their children, and so the child's attempts to bond with the caregiver are not as insistent and clear during the two minutes of "still face."

If a parent does not connect with their child and mirror their feelings, the child will have problems regulating stress and understanding their own and other people's feelings. For example, a child can be bewildered if they expresses stress or sadness to their caregiver and is then met with a smile. This child will create what the developmental psychologist Daniel Stern calls "a false self," where a person's inner needs are all suppressed. The false self puts a facade up to the world: how you look becomes more important than how you actually feel. The child loses the ability to read faces and connect.

"This makes a child unable to develop an expectation that others can feel like they do, especially in stressful situations," write Brandtzæg and Torsteinson in the book *Barn og relasjonsbrudd, Bind 2, Mikroseparasjoner* [Children and relationship breakdowns, Part 2: Microseparations], which was cowritten with psychology professor Lars Smith.

It creates lonely children—and lonely adults.

One characteristic of the lonely, as John Cacioppo's research shows, is that they are unable to read other people's faces as the non-lonely can: they do not become happy when they see smiling faces and are far more afraid of being disliked by others than the non-lonely. The lonely are less able to understand people and are unable to put themselves in someone else's place.

"When you combine experiencing much less joy from meeting nice people, with, in social situations, a limited perspective that involves a disproportionately strong focus on feeling threatened, be it real or imagined, the unfortunate result is poorer social abilities and responses, which in turn eventually reinforce the solitary person's isolation," writes Cacioppo.

In the story of Toffle, it becomes clear how much lonely people avoid eye contact; this is how Toffle tries to make himself invisible. Finding himself on the perimeter of a garden party with merry-go-rounds and lanterns in the trees, Toffle retreats into the shadows instead of entering the party and making new friends. "How lonely it must be to be a Toffle no-one sees. So who will comfort Toffle and explain the way things go? They'd know that he was there, if he would only say hello." Toffle is afraid. He does not venture toward the colors and fireworks, and this just compounds his loneliness. But it somehow feels easier like that.

I recognize this specific loneliness, and it's something that increased during the Covid pandemic. As the restrictions eased off and the "Finally, we can party!" invitations started landing in my inbox, my rejections became correspondingly frequent; I couldn't bear meeting all these new people. All the looks I had to deal with. Being seen. Being watched. Having to make eye contact. It was suddenly too intense, too invasive. I couldn't bear having to understand and keep up with what people thought of me. Because looks affect us, whether we want them to or not. Biologically speaking, there's something peculiar about human eyes: we can clearly see the white, the part surrounding the colored iris, and the reason for that is linked to how we communicate. Evolutionary psychologists believe that the visible white area is there because eye gaze has been and still is extremely important to us, and every single nuance of how the gaze falls, what and who we do and don't see, has been of decisive importance to us. In social situations, we are extremely concerned with where other people are *looking*, and we follow their gazes avidly. You notice it clearly at a big party: who looks at whom, who looks away. You notice it when you try and fail to make eye contact with an acquaintance. If you try joining a conversation, you will naturally start by attempting to catch the eye of someone in the group. But if nobody looks, your effort to socialize will fall flat. Not being seen, of course, has major consequences. The invisible are, by definition, not part of the herd; they are not those who should be rescued if something goes wrong. The invisible cannot expect to be cared for or loved, to be accepted and safe. The invisible can disappear, can be left to die. And when we don't have or can't use language, looks, facial expressions, and movements become even more important; and for children, they're crucial.

I'm reminded of this as I reach for my smartphone and my daughter says, "Put your phone away, Mom … Now, Mom! You said you were going to read for me!" She looks at me angrily and hands me the book. We snuggle up with the story of Toffle, and while reading it aloud one more time, I look down at her, and she looks up at me, her eyes full of life and curiosity, and in that moment, I feel an intense connection and sense of joy. It's one of the best feelings in the world, and one that I'm not alone in liking. To drown in the loving gaze of another person, be it a child, a lover, or a dear friend, is pure bliss.

Researchers have been examining this eye contact between parents and children in more depth. An experiment involving mothers and babies showed that when the mother looked directly at her child's face, it activated the same areas of the brain in both the parent and child as though they were singing in harmony. But if the mother looked at the baby from a slight angle, with her head partly turned away, there was a noticeably worse connection. It is the direct, open, face-to-face inquiry that really connects us. But the sense of joy I experienced when I had this connection with my child may also have something to do with a hormone: oxytocin.

While it may seem banal to talk about hormones when we want to learn more about attachment, oxytocin is a useful part of the story. Of course, it's not synonymous with love and care; it's just a small boundary stone.

The discovery of oxytocin in 1906 has made us better able to understand attachment. When Henry H. Dale first described the hormone, it was after he gave human oxytocin to a cat, which experienced uterine contractions as a result. One thing that we have in common with cats—with all mammals, in fact—is that we have oxytocin, and it does more than cause uterine contractions. Yes, it plays a central role in childbirth and breastfeeding, but researchers later discovered that it is also connected to attachment and caregiving. One experiment showed that mice, after getting a good dose of the hormone, could take particularly good care of baby mice, even if they weren't their own. Conversely, mice with low levels of oxytocin would behave indifferently toward their little mice pups, even if they *were* their own. Oxytocin levels in the body are affected when we orgasm, during childbirth and breastfeeding, by drugs, alcohol, and satiety (which may shed some light on the connection between loneliness and drug addiction and eating disorders, since food and drugs to some extent mimic the feeling of attachment). And, you've probably guessed it, oxytocin is connected to eye contact. Oxytocin makes us look for eye contact with other people, as an experiment conducted at the University of New South Wales in Sydney, Australia, has shown. Several experiments where the test subjects inhaled the hormone through a nasal spray while looking at pictures of faces have implied that the test subjects look more toward the eyes in the pictures. This presumably means that eye contact and attachment are closely linked. In another experiment, men who received oxytocin before interacting with their babies were more caring fathers— and this behavior appeared to be contagious: the father's open contact with the child caused a change in the child's behavior, even when no dose of oxytocin was administered. It consistently showed that if the father behaved in a certain way, the child behaved accordingly. And this is important: it means the child and the adult were mirroring each other.

One researcher has contributed more than anyone else to our understanding of how we connect with each other through eye contact. Vittorio Gallese is a professor of psychobiology at the University of Parma. In 1996, Gallese was studying a small primate called a macaque, in order to understand more about the human brain, when he discovered something strange. When the macaque did something, like walking over to a banana and peeling it, specific neurons in its brain would fire, and Gallese was able to measure this. But he also found that when the macaque just *watched* another monkey doing a certain thing, its brain neurons fired in the same way, as though it was doing the same thing itself. Gallese called these neurons "mirror neurons," and since then he has become an internationally celebrated and recognized interdisciplinary researcher. Gallese believes that these mirror neurons, which he observed in our smaller primate cousins, are the rudimentary origin of human empathy, our ability to understand how others feel and see the world from their point of view. Because we humans have mirror neurons too. We see what other people do and it automatically makes us feel like we are "walking in their shoes"; we experience the world from a slightly different angle than our own. When we look into each other's eyes, we become momentarily part of each other's reality.

Mirroring each other is something we do constantly. People we meet and relate to roam the corridors of our brains like neurological shadows: they tie their shoelaces, eat bread, and yawn, and our brains act as if we are doing the very same thing. We even *do* the same if it does happen to be a yawn; because yawning is so contagious, we mirror it immediately. We see someone in pain on television and grimace ourselves, as though we are the one suffering. We notice another person's mood, their tempo; we notice whether they are stressed or not, angry or happy. A very common amateur experiment involves standing among a group of people and folding your arms, and seeing how long it takes for everyone in the group to do the same. It's surprising how quickly it happens.

But when a friend is suffering, our mirror neurons fire even more. In an experiment, one of two people in a room was exposed to pain (via a small electric shock), and this affected the person watching the person in pain, but more so if the shocked person was a close friend—yes, when that was the case, the brain signals appeared to indicate that the person watching was in pain. Someone else's suffering is our suffering, and if it affects someone we really identify with, the pain becomes especially strong. When I see my child fall and scrape her knees on the asphalt, I feel her pain like a shudder through my body. Some of us will sob a little when a beloved character dies in a novel we're reading. Even when we're just listening to stories around a campfire, our brains are activated so that we can see the world from a new vantage point or from another time, through the eyes and fingertips of a person we may never have met.

We mirror when we are just tiny and unable to speak; we see someone bend over our pram, who smiles and waves, and we try to do the same in return. A mother gapes with her mouth to encourage her child to open up for a spoonful of baby food, and it works. Perhaps this is what Sigmund Freud was thinking about when he described "oceanic feeling" (a term he borrowed from the author and Nobel laureate Romain Rolland): when we mirror a dear friend or lover, as a baby might do when lying close to its mother, it feels as though we become one and the same person. "At the height of being in love the boundary between ego and object threatens to melt away. Against all the evidence of his senses, a man who is in love declares that 'I' and 'you' are one and is prepared to behave as if it were a fact," writes Freud in *Civilization and Its Discontents*. The human gaze is one such fusion: it is one of the gateways to love, to caring and learning to become a human being; it is one of the bridges between us. I remember looking into my daughter's eyes while I nursed her. It was as though nothing else existed.

We learn from each other; we are each other's parrots. My daughter is always playing copycat with me: "Stop copying me!" I say. "Stop copying me!" she replies delightedly. "I really don't want to play copycat!" I say. "I really don't want to play copycat," she replies, laughing, and I suddenly realize that this game of ours is the same one that I played with my dad forty years earlier. But even when it's not a game, I see how my daughter mirrors the expressions and movements that I am using. She even swears like me if she bumps her head, much to my dismay.

Even when we can't properly see what's going on around us, our so-called F5 mirror neurons become activated. Yes, even when we only *hear* that another person is doing something just out of range, we still clearly imagine what is happening. For example, if someone enters the kitchen and we hear the sound of running water hitting crockery, we'll imagine that the person has started washing up, though of course something else could be happening. But when we can only *see*, for example, film clips of people talking while the sound is muted, it activates a part of the brain associated with language comprehension called the Broca area. This area does *not* become activated when we see a barking dog, however, because we are constantly wanting to connect with other people, to know what they know, in order to understand and imitate them. We read each other through our eyes, and by doing so we become part of each other's inner world.

So it can be lonely to exist in a culture where you only rarely meet someone else's gaze. Imagine walking down a street where nobody makes eye contact. Imagine whether friendship would even be possible without eye contact. Ika Kaminka spent several years living in one such place, a country where eye contact is in short supply.

"You kind of get used to not having eye contact, and not getting a hug. Being a foreigner in Japan is tiring. You miss being looked at and having skin contact. I'm always craving touch when I arrive home from Japan," she says.

Kaminka translates Japanese literature into Norwegian and has among other things translated most of the books by the Japanese author Haruki Murakami. The effect that these years in Japan had on Kaminka was twofold: she fell in love with the country and its culture, and she was inspired into a career working with its language. But it wasn't easy. Japan sits at the top of the loneliness statistics—there is something about this culture that fosters loneliness.

"I had plenty of friends, and I got a boyfriend. But the lack of physical closeness meant that it could still feel quite lonely," she says.

In the West, looking directly into a person's eyes is associated with having nothing to hide. Looking away is a sign that you are concealing secrets and lies. But in Japan, making eye contact with a person is considered an intrusion and is therefore something you rarely do.

"When the first Westerners arrived in the country in the 1800s, they described the Japanese as untrustworthy, people who wouldn't meet their gaze," Kaminka says. "They called them 'shifty-eyed.' In the books I translate, it might say that one of the characters *looks directly* at the other. You perhaps won't notice if you're a Westerner, but it's actually a special moment; it indicates that the character crossed a boundary, that a special form of contact occurred, and it can be experienced as both uncomfortable and intimate."

Looking each other in the eye wasn't commonplace when she moved to Japan in 1986. She lived and worked there for four years straight, and soon learned that there are many sets of rules in the island kingdom. There are different rules for writing a letter in autumn than for writing one in spring. There are specific and complex rules for gift-wrapping and sword-sharpening. And there is an unspoken rule that you don't make eye contact with people, something Kaminka learned while learning the language.

"What I found in Japan was that it was easy to make mistakes, but Japanese people were also very patient with foreigners. It was easy to offend someone unintentionally—although some of the rules are stated and can therefore be learned. But I think it's much worse to come to Norway as a foreigner. Here, we pretend that we don't conform to any social rules. And that's of course not true," Kaminka points out.

Kaminka's open, Western, Norwegian demeanor had to be toned down in Japan. If she forgot where she was for a moment and looked directly at someone while walking down the street, they would recoil.

"They would either look away or seem totally perplexed," Kaminka says. "You might have a flustered Japanese person quickly apologizing for obstructing your view."

During the pandemic, I missed it for the first time: walking along the street and meeting the eyes of a stranger; smiling at a random woman with a pram on the bus; or looking directly at the people working in the shop. Wearing the face mask easily created a slight distance. I thought about how in Japan this was all very normal. In Haruki Murakami's novel *Sputnik Sweetheart*, one of the characters exclaims: "Why do people have to be this lonely? What's the point of it all? Millions of people in this world, all of them yearning, looking to others to satisfy them, yet isolating themselves. Why? Was the earth put here just to nourish human loneliness?"

But when it comes to loneliness, there is perhaps something quite special about Japan. Loneliness is spreading throughout the country: surveys show a poor work-life balance and that a huge number of Japanese people use technology to fulfill their social needs. The growing hordes of *hikikomori* have chosen a life without eye contact, a life on the internet. The word *hikikomori* simply means to withdraw and confine yourself indoors. These are young people in their twenties who never move out of their parents' homes. They don't get jobs. They don't go out into the world. They are either online or gaming, behind drawn curtains and closed doors, in boy's bedrooms (because this mainly concerns men), where they live a kind of eternal teenage life. However, many of the original hikikomori are now reaching middle age, and when their parents die, they are finding themselves alone, with no experience of the outside world.

"It's mostly about shame, but it's also about pride. Pride stops many of them asking for help," says psychiatrist Saitō Tamaki of the University of Tsukuba.

It was Tamaki who first described the phenomenon, as early as 1998, and has written a book that includes figures on the subject. To be called a hikikomori, you must have confined yourself to your house for at least six months and be free of any other diagnoses. Today, the Japanese authorities estimate that there are 1.1 million young people living this way, but Professor Tamaki believes that the number will increase to 10 million in the coming years. He believes it is related to the excessively performance-oriented culture they live in.

"They feel like they are failing at school or work," he explains. "Many people who isolate themselves have a very negative self-esteem. They feel unworthy compared to others. They judge themselves. It's a vicious circle. When these people withdraw, they may feel better in the short term. But in the longer term, they will be much worse off."

The demands of Japanese society are too great; there is no room for error. At the same time, there is a strong sense of honor among Japanese parents to always take care of their children, even if they are adults. Having said that, the number of hikikomori is rising in every country where there is a brutal performance culture: in South Korea, China, Taiwan, Spain, Italy, the United States. Here in Norway too. In all these countries, there are young people who totally retreat from the world and stay in their bedrooms. They don't get good grades, a career, or a partner. They become invisible. Their world becomes tiny and controllable. Their lives become reduced to staring at a screen.

When Toffle becomes lonely and afraid, it is easy to imagine him becoming a hikikomori. Toffle's world, of course, lacks the technological and economic conditions for that to happen, the conditions that today allow young men to stay at home, indoors, with their eyes fixed on the screen, never going out to find work and friendships. Withdrawn teens and adults *have* to go outside eventually, of course—at least, they did until recently, because over the last ten or fifteen years, it has become increasingly easy to do more of the things we need to do with our eyes and fingertips. What does this do to us, when we look more at a screen and less at each other? While Japan is at the extreme end of the scale, we are experiencing the same thing all over the Western world. The pandemic showed us that we don't even have to leave home to have meetings, sing in a choir, look up a sixteenth-century manuscript, or sit by the deathbed of a family member.

That we had these digital solutions linking us across the geographical divides was on one hand a blessing, but on the other a curse, because they are so bad at facilitating real human contact. I have been thinking about how this affected us: how social media and video conferencing never enables us to make genuine eye contact, how it often leaves us exhausted and unsatisfied. It lacks the ability to genuinely connect us, which perhaps explains why talking via Zoom or Teams can feel so laborious: if you want to give the person you are talking to the illusion of looking into your eyes, you need to look at the camera lens—but all you see then is a tiny green light. Trying to catch the other person's eyes via the screen means looking away from the camera and thus not into the person's eyes. This constant pursuit of eye contact may be one reason why I felt so worn out and frustrated during the pandemic. But soon the tech gurus will solve this problem for us, and in a few years we will be able to look perfectly into each other's eyes. But how will that then change us, when we don't have to go outside to meet people physically, when the human gaze can be replicated by pixels on a screen?

The human gaze has already become part of one of the world's largest economies: the attention economy. In the digital world, our eyes are worth their weight in gold. What we look at, when we look at it, how long we look at it for, whether we click "like" or "heart" or nothing at all—all these tiny actions generate money for Silicon Valley's tech entrepreneurs. Elon Musk bought Twitter for $44 billion in 2022, which speaks volumes about what our clicks and glances are worth. What drives social media, and one of the reasons we get so hooked on it, is the strong feelings that govern it. We humans love strong feelings; the stronger the feelings, the more we click. The faster we click, the shorter our attention span becomes, the more fragmented the world around us seems—and the more we are rewarded in the form of clicks and thumbs and hearts. The unpredictability of the clicks we receive makes us hungry for more, like the experiment where lab mice were rewarded only sporadically and never knew when food would fall into the cage; it made them desperate for more. Regular and predictable rewarding, on the other hand, made them calm and happy.

I never know if I'm being seen and acknowledged online, or quietly ignored or overlooked, and this often stresses me out long before I reach for my phone to check how many thumbs I've received for a post or photo. This is the nature of social media and the addiction it creates, driven by unpredictability and the hunger to be seen; by the confirmations and rejections; by the looking at others and our desire to imitate them, to envy the lives they live as they show off their successes.

"When we stop looking, the social media platforms dissolve into nothing," writes author and journalist Lena Lindgren in her 2021 book *Ekko: Et essay om algoritmer og begjær* [Echo: An essay on algorithms and desire]. "I increasingly view our senses as natural resources in themselves.

"This enormous industry is controlled by visual capitalism," she says, while describing the shift in reality she has felt after using social media for too long. Online, the pauses and doubt, the friction and inertia you experience when meeting a person in real life aren't there—the feeling of sitting with another person in the same room, the glances you meet before looking elsewhere. So much of the internet is governed by a nonphysical experience, by the eye alone, that it moves you away from reality.

"Reality is a physical experience, and the brain coordinates this via touching and seeing and interpreting other people," Lindgren tells me. "That's how we became human! There's a big difference between communicating via a screen and meeting a person physically; our senses know the difference. The world's richest companies use our sense as a commodity, and they've become that rich because they are controlling where we look."

Within these platforms, there are many people who see others as no more than digital phantoms that flutter across the screen. "I don't meet people," wrote one of my Facebook friends recently. "I feel invisible, and I can't bear going out. And that makes me even more invisible."

So begins a negative spiral, from just a glance: the invisible find themselves totally connected to the internet and totally disconnected from physical contact with flesh-and-blood people. Of course, being so insignificant that you're not seen, that you retreat into the shadows, meets all the criteria for loneliness: you are not part of the pack, not really. You are in constant danger. And the human gaze is one of the gateways to being understood and seen and looked after. If another person's face is a mirror that causes our mirror neurons to react, and if another person's attention is so crucial that we're afraid of losing it, it's no wonder the human face and eyes have such great power. The internet only amplifies this. And the research shows that if we sit for extended periods with our eyes on the screen, it undermines our ability to read faces, and this in turn creates loneliness, which in turn makes it even more difficult to read faces, and the negative spiral of withdrawal will have begun.

"Now a university has started a course in reading faces, because the students have been using screens too much," points out author Noreena Hertz, who has written the book *The Lonely Century*. "This means of 'mass distraction,' the smartphones, creates a meaner and crueler society: people smile less when they have a phone, and that's not dependent on them using it or not, what matters is that they have it with them."

In efficient cities, distracted by our smartphones, we refrain from initiating social contact, we make each other invisible. As the Swedish novelist and playwright P. O. Enquist observes: "A human being can live without sight; the blind too are human. But if one is not seen, then one is nothing."

One of the founders of modern psychology, William James, also recognized this basic need to be seen, and what happens when we retreat into the shadows, as Toffle did. "No more fiendish punishment could be devised, were such a thing physically possible, than one should be turned loose in society and remain absolutely unnoticed by all the members thereof," James wrote in 1890 in *The Principles of Psychology*. "If no one turned around when we entered, answered when we spoke, or minded what we did, but if every person we met 'cut us dead,' and acted as if we were non-existing things, a kind of rage and impotent despair would before long well up in us, from which the cruelest bodily torture would be a relief."

It is easy to think that a look is unimportant. It can seem so harmless. But next time you're at a party, on a bus, in a lecture hall or a classroom, try to notice where the eyes are looking.

Who's being looked at, who's being overlooked? Who is attracting everyone's friendly attention, who is disappearing into the background, who is being consciously or unconsciously ignored? To deliberately make someone invisible is to exert a subtle form of power.

In her book *Hersketeknikker* [Master suppression techniques], Norwegian journalist and debate moderator Sigrid Sollund describes how one government minister who was invited onto her radio show avoided eye contact with both her and his debating opponent. The minister looked the other way and out the studio window, and by doing so left his opponent stunned. She reacted by becoming *prosocial*, that is, she tried extremely hard to get his attention.

"The other debater went from being critical, to unsure, to then praising the ministry's efforts," Sollund writes. "Anything to get his attention, eye contact or acknowledgment that she was sitting there at all. I found it so uncomfortable that I just wanted to end the session as quickly as possible. The master suppression technique worked incredibly well."

In her book, Sollund names both excessive staring and looking at a point on the forehead of the person you are talking to as ways that can easily alter the power dynamic. They are effective ways of making someone flustered, because they are so vague and difficult to accuse someone of doing. It was just a *look*, after all: "I can claim that you looked strangely at me until I'm blue in the face. But as long as you deny it, I've got no case," she points out.

A look, or a non-look, can sometimes imply direct rejection, or *ostracism*. In recent decades, ostracism has been the subject of a lot of research, and this research shows there is a link between social exclusion and the degree of warmth and closeness in the relationships you have, as well as the knowledge and material assets you get access to in life. People who experience persistent exclusion are at greater risk of attempting suicide, of depression, and of committing violent acts such as mass shootings.

The word *ostracism* comes from the ancient Greek word *ostrakon*, which was used to describe a shard of pottery. Greek citizens wrote the name of a person they would like to have removed on a pottery shard as a means of collecting votes; if the number of votes reached six thousand, the person concerned would be exiled for ten years. It is a very brutal loneliness: the terrifying certainty that six thousand people will banish you from the community. After all, Western culture is actually founded on a story about ostracism: Adam and Eve are living happily in paradise. But one mistake, one bite of the wrong fruit, results in God banishing the first humans from their Garden of Eden, forcing them to forever live as outcasts, and in harsh conditions, because they also know, by just looking at each other, that they must be ashamed of being naked, and are no longer in paradise. Just one glance can turn you into Toffle, hiding in his house.

And I know that one glance has this power because of a very interesting experiment I read about, which deals with glances and ostracism. It is based around a game called Cyberball and has helped scientists understand more about what happens when we feel overlooked and excluded. In one version of Cyberball, a team of devious neurologists sets up a somewhat cruel experiment for three participants, who are asked to play a supposedly multiplayer computer game where they seem to be throwing a ball to one another in turn. After a while, the ball stops coming to them—at least, that's what each of the players (who cannot see one another) thinks. They are in fact being treated alike; none of them have been thrown the ball, and each player believes they have been rejected by the other two. All three participants start feeling quite unhappy, even though it's just an experiment in playing a fictional game on a screen against people they don't know. During the game, all three participants lie inside their own fMRI machine, which allows the researchers to scan their brains as they experience the rejection of not being given the ball.

What the experiment's creators found was that when this kind of rejection is felt, the brain is activated just as it is when we are in physical pain. Rejection is felt like a slap in the face. And this sense of exclusion generates three clear reactions: One is that you become far too socially active—prosocial—and try to regain the attention of those wanting to push you into invisibility. Another involves you withdrawing, either because you can't bear having to struggle for their attention or because you feel sure that there are other social opportunities out there. The third reaction is to become aggressive. All three reactions are responses to loneliness.

The participants feel rejected even when fully aware that the game is computer-driven and pre-programmed, or when the "ball" is instead presented as a bomb that could kill them at any moment. *Even then*, their feeling of exclusion is more dominant than their fear of the bomb. We humans are hypersensitive to exclusion, significantly more than we are to the opposite. We notice rejection far more than we do inclusion, the reason being that it is safer to notice every little sign of exclusion—so that we can adjust and end up within the community again—than it is to be too trusting and inadvertently find yourself being thrown out of the pack. The fear of exclusion is, as we know, a strong driving force: it makes us adjust back into the community and "normality." Feeling lonely is like hunger, a deeply unpleasant, physically stressful experience that forces us to return to the pack and do what it takes to fit in.

In 2014, researchers investigating the theory of how strong the effect of ostracism is wanted to see if the strange and unpleasant Cyberball game can be played *solely by eye movement*. They held the "invisible man" games, where the players had to exchange glances, just as they had previously exchanged a digital ball. During the game, one of the participants was excluded; the other two continued to send glances back and forth, while the third player received no glances at all. And it turned out that this was very uncomfortable for the excluded player. It even worsened the player's cognitive abilities, because their concentration was focused on trying to be accepted by the others.

"Being invisible to others will be experienced as exclusion, and this kind of all-encompassing exclusion really has something of the same effect as when one exchanges objects," writes the German research group, who found the experiment to be highly successful: the participants felt real social pain when they were excluded by a glance.

It's not strange that we hide and feel ashamed of everything that might lead to rejection, if what we're also doing is trying to protect ourselves from pain! And not being seen, both literally and figuratively, triggers pain.

The researcher Kipling D. Williams at Purdue University in West Lafayette, Indiana, has written a book about ostracism and created a model for understanding rejection that he calls the "need-threat" model: basic needs are threatened when you are actively pushed out of a community.

"In the long term, if people are exposed to repeated or long-lasting episodes of ostracism, their attempts to restore what they have lost will eventually give way to feelings of alienation, helplessness, depression, and despair," he writes.

Williams describes how the social pain caused by ostracism occurs in several stages. First, the individual's critical need to belong and have control leads to intense attempts to reenter the community. Then comes withdrawal and reflection, which trigger several strategies for getting back into the community, which is then followed by resignation, if nothing works. In other words, he describes the different stages of loneliness and its reactive feeling, depression.

But when you really need Harry Potter's invisibility cloak is when other people's looks become too invasive and aggressive—when their eyes are full of scorn and contempt, and you want to sink to the floor and disappear. Or when people stare at you in disbelief, to show you that you don't belong to the norm, you stand out from the pack, you are weird. As a rule, the bodies of those considered different are assigned characteristics like stupid, lazy, unimportant, hypersexual, unrestrained, aggressive, pathetic, mentally incompetent—abnormal. A simple glance from someone, at a person's appearance, can decide whether that person will receive harsher punishments or poorer health care, or whether a rape or murder will be properly investigated.

"We can never escape the prejudiced gaze," says Norwegian author and activist Guro Sibeko about how she experiences racism. Among other things, she has written the 2019 anti-racist book *Rasismens poetikk* [The poetics of racism]. Sibeko's dark skin is visible, whatever time of day it is. She can find herself being stared at when she least expects it. And that is merely a prelude to the unpleasant comments, the exclusion, and the violence.

Ali Esbati, an Iranian-Swedish politician and social commentator, points out the same thing. And he begins with children just playing innocently in a public space.

"Try looking for it yourself, at how white and blond children are treated differently to those with dark skin or hair. Read the silent language of looks, facial expressions and body language," writes Esbati, a survivor of the Utøya terrorist attack, in his book *Etter rosetogene* [After the rose parade]. "I'll tell you what I see. Not always of course. But often enough for it to become logged in the operating system. For the first category, patient smiles, facial expressions and body language that says 'oh, aren't they cute.' For the other category, a different message. Wordless, but clear. 'Oh, what annoying kids.' 'Can't the parents teach them how to behave?' 'There's too many of them here now.'"

It's a seemingly invisible form of exclusion, which affects children from a very young age. But the exclusionary gaze is not just a racist one. Anyone with a body that isn't defined as "normal" risks becoming a thing to be stared at. It affects anyone who looks different, the wrong different, who is fat, has a disability, or appears to have a skin disease or extra body hair, all the "monsters" and "freaks." All good reasons to stare used to be a financial opportunity; yes, in our part of the world people happily paid money to stare at unusual bodies at freak shows and in seedy circus tents. Even Scandinavia's indigenous Sami people were put on display. And it's probably no better now; it's simply been moved over to our screens. Now we can stare freely at those we perceive to be different from our own living rooms.

"The advent of photography in 1839 began to shift much staring at the luridly different bodies of freaks from the actual public encounter to the private Victorian photo album, ephemeral media," writes professor Rosemarie Garland-Thomson of Emory University, who works on "disability studies" and feminism.

The unusual body is stared at because it doesn't blend in with the crowd or fit in with the pack, and we have a need to categorize it. The unusual body is perhaps also "the weak body";

it belongs to people who may need extra protection and care, and so those who have these unusual bodies are even more vulnerable to the exclusion and alienation that results from the staring.

"If staring is the effort to make sense of the inexplicable, to craft a narrative of recognition from incoherence, then the target of staring is often that which seems strange or unfamiliar," Garland-Thomson notes.

The staring isn't about wanting to help or include. Staring creates distance between people. Professor and author Jan Grue is part of Norway's largest minority: those who live with disabilities. And he knows a lot about what the ostracizing look feels like. Grue has written two books about what it's like to be a wheelchair user. One of them, *I Live a Life Like Yours*, was even nominated for the Nordic Council Literature Prize and won the Norwegian Critics Prize for Literature in 2018. In his books, Grue describes how people manage to avoid looking at him or talking to him, or will just assume that he is neither an adult nor mentally competent. Instead of him, they look at and talk to his wife.

For example, eye contact is crucial at an airport when busy staff are wheeling him out of sight in one of the airport's basic wheelchairs, while his highly advanced electric Permobil is being loaded into the plane's cargo hold. "It's important, in these situations, to speak up early on. To catch eyes, to speak loudly and clearly. To show that you are sane, that is, to disprove the assumption that you are not," he writes in *Hvis jeg faller* [If I fall].

He also describes the looks that follow him and his wheelchair when he must lecture or stand on a stage, because he is, after all, a professor and author. He must catch people's attention; in his profession, being invisible isn't an option.

"The moment I go out onto a stage is the moment I become most aware of being a wheelchair user. I know I'm an unusual sight up there, I feel the weight of the looks I usually ignore," he writes of the staring.

"And as every person with a visible disability knows intimately, managing, deflecting, resisting, or renouncing that stare is part of the daily business of life," writes Professor Garland-Thomson. She believes that a look, the stare, can lead to different forms of exclusion, and from there to loneliness. And when you measure loneliness, people with disabilities rank significantly higher in the statistics than most of the population. They are discriminated against in working life; they are isolated. People with disabilities are socially excluded both privately and at work. They are gawped at. And they suffer violence and rape more often than the general population. But the loneliest members of society are probably those with visual impairments. In a survey from 2018, half of this group said that they experience loneliness. Have I, too, created a sense of exclusion by writing an entire chapter about the human gaze in this book—if you, the person now reading it, are blind?

Worse than the invisibility, worse than the staring, is the threatening look. This is a direct and undisguised promise of exclusion from the community. It is a warning of violence. The physician and researcher Anna Luise Kirkengen believes it is even more powerful than that, claiming that a threatening look is pure violence. Her book How Abused Children Become Unhealthy Adults is based on research into how we are affected by abusive relationships—and it shows, of course, that the earlier in life we experience these abuses, the more harmful they will be to the body in the form of toxic stress. Children who live with violence and threats of violence are vulnerable to developing a range of symptoms and diseases later, as adults. The connection between toxic stress—which arises from chronic loneliness—and physical pain and suffering is exactly what Julianne Holt-Lunstad documented so thoroughly with her research from 2010, where she found that the most dangerous thing for the body—more dangerous than daily cigarette smoking—is being lonely. Dr. Kirkengen, however, had already discovered this connection in the 1990s, while writing her PhD thesis. Her work on it began when she was baffled by a patient who, over the years, came to her suffering from a string of vague ailments. It was only when the patient was murdered by her husband that Kirkengen realized that what had been a mystery in the medical sense was obvious when looked at from a totally different angle. The woman who had visited Kirkengen's office had been subjected to regular violence and threats from her next of kin. This meant that she never felt safe or experienced any sense of belonging in the one place that really should have shielded her from loneliness—her relationship. Kirkengen then began to investigate the link between violence and physical ailments. Today, she is professor emeritus in social medicine at the Norwegian University of Science and Technology and takes hateful looks very seriously—because the fear of violent exclusion, the most terrifying form of loneliness, can be triggered by just a glance. It is the petrifying gaze of Medusa.

"To be the object of someone else's disrespect, to experience rejection, hatred and objectification, is damaging to one's self-esteem," she says. "And someone's contempt can be deadly: actively, in that you could be killed, or in that it leads you towards drug addiction and risk-seeking behavior, self-harm and suicide. It is meaningless to distinguish between different types of violence. If you get this warning look, you have already become an object, you are on your toes. There is already violence in that look."

The look that warns of impending violence is extremely scary. One look can trigger a stress response and force the victims of this violence to take protective measures, which in turn makes them ill from stress. And it is an experience that remains in the body. Today, more than thirty years later, Martin Eia-Revheim, a well-known Oslo-based entrepreneur, still remembers the way his violent and alcoholic father looked at him. "Dad taught me the look of punishment. He would give me this stare, in full public view, to show me that I had crossed the line, that there would now be a punishment—or *reprisals*, as he called it when we were alone," he writes in his autobiography *Å sette sammen bitene* [Putting the pieces together].

Violence is the most extreme form of exclusion from a community and is closely linked to the feeling of loneliness. It could be that violence is what we fear most when we feel lonely. Or perhaps loneliness is vaguer than that. It's not so much a pronounced fear of violence as it is a vague discomfort, a knot in the stomach, a sense of unease and anxiety. The gaze is perhaps one of the most basic ways of excluding someone, a decisive piece in the game, when it comes to both inclusion and exclusion. The gaze is often how we first encounter another person; we exchange glances long before exchanging words, and even more so now, in today's cities, than when we lived in small, tight-knit communities where everyone knew each other. In the city, we meet new people all the time, and we meet them with our eyes and faces.

So much effort goes to attracting attention; we are so afraid of not being seen or being seen the wrong way. The book you are reading now is also an attempt to be noticed: I want so much to be finally seen, for the outline of me to become clear.

The plume of smoke that rose from downtown Oslo on July 22, 2011, was someone wanting to be seen, as was the manifesto sent to 1,003 email addresses. So what is the most gruesome revenge we can inflict on a man who did the unthinkable, who planned and carried out a terrorist attack that resulted in the deaths of seventy-seven people, most of them children and teenagers? The poet Cecilie Løveid outlines a very special punishment that involves, among other things, not being looked in the eye, not being documented or made visible on the internet:

> As we know, he is condemned to look everyone in the eye
> and as we know no one wants to look him in the eye.
> The bottomless reservoir of self-defense
> will be reversed.
> Sudden, brave, unexpected questions will
> lay him bare.
> He will not be allowed to change his name. Someone will
> say his name and he will turn.
> He will become the famous penitent. He must carry
> out this task alone, without the use of the internet,
> manifestos, puppets, stand-ins or doppelgangers.
> The reasons of judgment also stipulate:
> That none of this penance can be photographed, sketched,
> cartooned or filmed.
> No audio recording can be taken, only the word *sorry* can
> be recorded.
> One cannot go back and change something that was felt.
> Therefore, and even if he does all this, there will be silence.

ONE EVENING, WHILE the rain outside lashes the windows, my daughter and I are once again hunched over *Who Will Comfort Toffle?* when she asks me why I look so unhappy.

"Because I know how Toffle feels. I know what it's like to hear the Groke howl," I say, hugging her even more tightly. I know what it's like to be invisible in a schoolyard.

I think about the people whom I have made invisible by not looking at them. Those I've met on the street, those I've hurried past indifferently on the train, those I didn't notice when I helped at the school fundraiser. The parents of my daughter's classmates I failed to smile at. All the times I've looked at my phone instead of the people I'm with. How I've crept past the beggar outside the local shop while deliberately trying to avoid making eye contact. All the times the face mask I've been wearing has obscured my expressions and made my eyes look evasive.

I take our book of collected Moomin stories off the shelf again—children love repetition, it's such a safe bet—and open it up at "The Invisible Child." While I'm reading, my daughter lies in the crook of my arm, snuggled into my armpit with her head resting on my shoulder. She occasionally looks up at me, before her eyes return to the pictures and her restless fingers settle around the edge of the book.

"Last Friday one couldn't catch sight of her at all. The lady gave her away to me and said she really couldn't take care of relatives she couldn't even see," Too-Ticky tells the Moomin family.

My daughter laughs at the strange scenario: an invisible child! To her it's just comical.

We read about how the Moomins succeed in making Ninny visible again. They simply allow her to live in an ordinary, chaotic, and loving family who clean mushrooms together on autumn evenings. They give her a safe bedroom, sew her a dress, hang a little bell on her, and make her some medicine according to a recipe Moominmamma finds among her grandmother's notes. They teach her games, because Ninny has never played before; initially, she is far too polite and orderly and quiet.

At the end of the story, Ninny protects Moominmamma from being pushed from a jetty into the sea. The little girl suddenly musters all her strength to defend someone she loves—and becomes visible.

"'I see her, I see her!' shouted Moomintroll. 'She's sweet.'"

"But *I'm* not sweet, Mom," my daughter says confidently.

I look at her and laugh. "Oh no?"

"No—I'm cool. I'm *tough*."

The whole bedroom is now bathed in the blue shadows of dusk. I rise to my feet and notice my tough little girl looking down in the mouth; she's not that tough anymore.

"Don't go, Mom! I'm afraid of the dark! Can you sit here with me? *I need to see you*."

I sit with her until her eyes are completely shut and the sound of her gentle breathing fills the room. With my hand upon hers I feel totally calm and safe for a moment, even though the night outside is so dark, and the pandemic has only just begun.

3

Why having
a good laugh
doesn't always
prolong your life

FEW THINGS MAKE me happier than laughing with my daughter. Sometimes she'll have a laughing fit where she throws her head back, holds her stomach, and shouts, "Mom, I'm gonna pee myself!" This shared laughter creates an intense feeling of belonging (and an intense feeling that you have to pee). I remember it being perhaps the most wonderful thing that happened to me as a child—that I made my father laugh. Whenever I laugh like that with my daughter, I picture my father and me, and how the laughter was like our own little island, an oasis of belonging and joy. A place where we were totally safe.

Humor is a basic part of interaction and communication; laughing calms the nervous system and helps us become friends—which is why mean and exclusionary laughter is correspondingly unpleasant. Originally, laughter had two purposes, and just like the human gaze, it can both include and exclude.

Laughing and smiling connect us when we don't have language. A baby, for example, will smile at its parents from the age of two months; and at three to four months, it will start to laugh, creating a bond and making communicating easier long before the child has developed language. In other words, laughter creates unity during the strenuous period of infancy and child-rearing. And we need it, too, since humans have the longest childhoods of all animals, in relation to life span. That we shifted from a die-young strategy to slow development and long life spans has given humankind one of its great evolutionary advantages. And it was due to the organ we have invested the most in: the brain.

The human body has always evolved as far as physically possible, because humans walk upright and therefore need narrow hips, while also needing to give birth to babies with large brains. The development of the mother's hips in relation to the child's head constitutes an important tipping point in human evolution. If the baby's head gets any bigger, the child will not come out; but if the mother's hips become wider, she will no longer be able to walk properly on two legs. Babies must therefore come into the world while they are still under construction, with an unfinished brain and body. If you compare a baby lamb, which will immediately get up and stagger about by its mother after being born, with a baby human, which will spend an entire year learning to walk, it's clear why having a close parent-child relationship is so important. A human parent must want to take care of their young for years, and any caregiver must be extremely attentive and caring while the child's brain develops and matures. The highly complex human brain is not fully developed until we are well into our twenties, so not only the parents but everyone around the child must have a strong sense of unity if they are to look after a newborn, vulnerable, language-less being who at first can neither walk, seek protection, nor find their own food. Our rather long life span ensures us a group of caregivers with a wide variety of ages. And this is where laughter comes into play. It enables us to communicate despite our differences.

When they laugh, babies produce higher levels of the afore-mentioned bonding hormone oxytocin. In an experiment using hospital clowns, German researchers found that when children laughed at the clowns, it lowered their anxiety and raised their levels of oxytocin, which were measured in the children's saliva after the show. And Dutch researchers found that when child-less women were given oxytocin, their reactions to the sound of children's laughter became the same as a mother's reaction to her own child's laughter. Oxytocin makes new mothers more empathic; they develop more altruistic traits, become better at putting themselves in someone else's shoes, and deal with crying babies in a more calm and relaxed manner. A child's smiling and laughter strengthen the bond with their parents and are consid-ered an evolutionary advantage. Laughter makes the challenges of parenting easier to face and helps the love to grow between parents and children. But the connection we get from laughter and smiling didn't evolve as a luxury. For the human groups who lived at the mercy of nature, strong bonds could be abso-lutely essential to their survival.

Evolutionary psychologists believe that the source of the primordial laugh is tickling, which can be done to the point of laughter to dogs, rats, and monkeys. This laughter probably orig-inated from both tickling and cuddling. With the development of language, humor has become more verbal and abstract. Yet this laughter has something animalistic about it: researchers have found that we can easily hear the difference between spontaneous and non-spontaneous laughter; even when recorded laughter was slowed down and the pitch became deeper, the listener could tell which of the two was non-spontaneous. But what listeners to these strange recordings couldn't do was distinguish between ape sounds and *spontaneous human laughter,* which shows that laughter in its purest form is close to our natural origins among the ape family.

Laughter unifies a larger group than simply those you can look in the eye or touch. While apes live in smaller groups and bond through grooming and cuddling, humans are organized into communities far too big to be maintained by physical contact alone; there are too many complex relationships to nurture for that to be possible. The average person has 150 acquaintances, a circle that contains both friends and close family, as the research of anthropologist Robin Dunbar shows.

"The results show that the degree of shared appreciation for both sets of stimuli had a positive effect on Affiliation; only humorous stimuli had an effect on Altruism," write Oliver Scott Curry and Robin Dunbar of the University of Oxford, stating that humor and altruism are strongly related.

Dunbar and his colleagues believe that laughter's contagiousness says something about how it works as a social glue, in much the same way as physical contact. Just think about it: when you've gone to the cinema alone and seen a funny film, or when you're at a stand-up comedy show and the whole audience bursts out laughing, it's impossible not to smile yourself. This is because laughter has a very physical effect on us, much like having your arm stroked or getting a warm hug, without anyone having to be near us. It can bring many people together in one spontaneous reaction.

Laughter is more of a social lubricant than a measure of how funny something is. After all, we'll laugh at something half-funny a close friend or a friendly neighbor says to us, just to make them feel good. Laughing is something we do on average eighteen times a day—rarely at jokes—and we are thirty times more likely to do so when we are in a group than when we are alone.

"Laughter appears to stand in need of an echo," wrote the interwar philosopher and Nobel laureate Henri Bergson. "Our laughter is always the laughter of a group. It may, perchance, have happened to you, when seated in a railway carriage or at table d'hôte, to hear travellers relating to one another stories which must have been comic to them, for they laughed heartily. Had you been one of their company, you would have laughed like them; but, as you were not, you had no desire whatever to do so."

Bergson points out that the apparent spontaneity of laughter hides the fact that it feels like something being shared by members of a secret club. There's a kind of camaraderie and complicity in it: "How often has it been said that the fuller the theatre, the more uncontrolled the laughter of the audience!"

This shared understanding is a powerful deterrent against loneliness and one path toward friendship and togetherness. Laughter is the expression of an underlying emotion: we show our amusement through laughter, just as we show anger by glaring. All these physical signals are given *to* someone. It requires an effort to laugh when you're alone, but very little when you're in other people's company, which is why TV companies use canned laughter to give the impression that their comedy shows are genuinely funny. Laughter is contagious, even when it's only a recording.

Research shows that at the age of eleven or twelve our friends suddenly become far more important to us, and that a central part of making friends is our ability to laugh together. Friendship also contains elements of deep seriousness and self-disclosure, and for a true friendship to emerge you have to risk showing a little of your own vulnerability: the risk of revealing your secrets and innermost thoughts is the quantum leap required when building a friendship.

"But when we focus too much on that, we're neglecting the value of joking around with one another and seeing what's going on with each other," says researcher Jeff Hall, who has investigated the conditions for becoming friends. "It's not that self-disclosure doesn't matter. It is that other things do, too." Hall believes that both friendly banter and great seriousness are equally important.

Researchers who have studied the importance of laughter in friendships have divided humor into four different types and have analyzed each one: whether you use self-defeating humor or assertive humor, whether it relieves social tensions and creates bonds or is downright mean and ridiculing. And although the researchers could see in twin studies that 14 to 25 percent of our humor style is determined genetically, the remaining 75 to 86 percent is shaped by our environment.

"Friendship dyads may afford young people a safe and secure context within which to observe and learn socially effective and successful humor styles," say researchers Simon C. Hunter, Claire L. Fox, and Siân E. Jones, who looked at the friendships and humor of children aged between eleven and thirteen.

The results of this study, which followed eighty-seven pairs of best friends over a period of six months, were clear: the inclusive, bonding type of humor prevailed over time and was "infectious" within each pair of friends, whereas the mean and ridiculing humor decreased. The social bond we create when we laugh is extremely important to us. So inclusive humor wins, and we are shaped by the humor of those closest to us.

"These results have clear implications for theories of humor style development, highlighting an important role for affiliative humor within stable friendship dyads," write the researchers behind the study.

Laughing together, at the same time, at the same joke, creates an almost incomparable type of joy and togetherness. It's a feeling I get whenever I laugh with my daughter, and one I remember from laughing with my father as a child. I knew he was my friend simply because he was laughing so uncontrollably with me.

"This emotion is accompanied by a series of biochemical changes in the brain, autonomic nervous system and endocrine system, involving a variety of molecules, including neurotransmitters, hormones, opioids, and neuropeptides," writes Canadian psychology professor Rod A. Martin in his 2006 book *The Psychology of Humor.*

What Martin describes is how reward hormones such as serotonin and dopamine sail around the brain when we laugh, creating a sense of well-being and relaxing the body—in other words, what happens when our *parasympathetic* system is activated. When this happens, the body becomes more attuned to playing, belonging, and friendship than it is to escaping and being aggressive, i.e., the opposite of the *sympathetic* stress responses. The autonomic nervous system includes two basic divisions: the first, the sympathetic system, ensures quick survival; but when we activate the second, the parasympathetic system, our pulse rate settles, we raise our heads, look up, ready for eye contact, with our ears maximally tuned for listening to human voices. The parasympathetic is known to researchers as the "vagal brake," due to its main component being the vagus nerve. This long nerve extends from the brain through the throat and past the vocal cords—and it is stimulated by singing and laughter.

"Narrowing of the larynx also modulates our voice to make it sound appealing or soothing—that is, to make it activate some-one else's vagal brake," researcher Nils Eide-Midtsand writes. "The task of this peculiar system is for mammals to signal to each other 'I mean no harm, you can safely approach.'"

When our *parasympathicus* system is activated, we nurture friendships and creativity, we tell stories around the campfire, we're more cuddly, our sex drive increases; we can also become prone to aimlessly sniffing flowers and loud fits of laughter, which means that parasympathicus is the network of attachment and the nervous system of laughter. When we laugh, the body relaxes and becomes full of happy hormones; we feel a sense of belonging and joy. And when we feel belonging and joy, we open up to other people, and we become more attuned to hearing their perspectives.

Researchers working on pedagogy at Wayne State University School of Medicine in Detroit, Michigan, believe that humor can be a very useful tool when teaching. If a teacher can appear falli-ble and self-deprecating, the resulting humor will make it easier for students and pupils to relate to the material with a high degree of empathy and interest. The whole body is receptive to information when we laugh. There are even psychologists who deliberately use laughter as a tool during therapy, because it creates the trust and fellowship required when sharing painful memories and experiences.

"In this context, humour created nearness, made the therapist accessible and authentic, and made the patient relax before the daunting task of opening up before a stranger," writes Patrizia Amici, a therapist in Bergamo, Italy, who deliberately uses humor when working with clients. The US-based Association for Applied and Therapeutic Humor defines laughter therapy as "any intervention that promotes health and wellness by stimulating a playful discovery, expression or appreciation of the absurdity or incongruity of life's situation."

For palliative care nurses, humor is something that helps them through the care they must give to dying patients. "Humor eases the tension and increases the emotional support perceived by both the nurse and the patient," writes a research team led by Inês Robalo Nunes at the Hospital de Santa Maria in Lisbon, which has investigated the use of gallows humor among palliative care nurses. These nurses work only with the terminally ill, and when their patients die, they usually mourn their passing without any acknowledgment or natural place for it, despite being so close to them at the time of death. Humor is of great importance to their ability to handle the emotional stress they are exposed to at work.

Laughing when terrible things happen dulls the pain. And while laughing doesn't seem logical when you're wracked by grief and sorrow, it actually is. Laughing with others is one way out of loneliness. And when my dad died, it was something I realized I had to try to do.

My dad would laugh often, and loudly. I clearly remember the first time I made him do it and the special bond we felt when laughing together. I don't quite remember exactly what we laughed about, just the strong sense of togetherness where for a few minutes we were totally safe. And like my father, I have always loved laughing with others; sharing laughter is like having a warm blanket wrapped around me.

He was a professor of philosophy, and his first philosophical treatise was about Henri Bergson (the same Bergson I quoted earlier) and his thoughts on the nature of laughter. When I made a speech at my dad's funeral, I talked about how he used to collect jokes. Many of his friends were present in church that day, people he once laughed with who were now laughing out loud as I told them about my strange and funny dad, and it somehow helped. Grief is so lonely, there's so much those around me cannot possibly understand; only he and I knew our history and could know the truth about our relationship, and now he is gone. But it helped to laugh. It was as if some of the darkest feelings around his death instantly disappeared.

A joke is a sudden and surprising and slightly dangerous comment that induces the kind of liberating laughter that can break through the walls we build around us. In the great and uncontrolled darkness that constitutes grief, there are ways to gain control and community.

"Lots of jokes are about death. They're a kind of fuck-you to the universe and the fact that we all have to die. And a lot of humor concerns our bodily functions, and how we are slaves to our own bodies," says comedian and comedy writer Markus Johansen, who has written the book *Hva er humor* [What is humor?].

For him, humor is about testing the moral boundaries of a society. We laugh because the boundary we have crossed is not as dangerous as the boundaries that come afterward; if we ridicule another person, then the laughter is a sign that they have crossed the line of society's norms. If people laugh, the one doing the ridiculing and the audience agree that that person is too far outside the norm. Humor then functions as an appeal for the person to return to society. It gives us an opportunity to break the moral rules in a safe way, just as play-fighting is a safe way to test the limits of violence. Or just as horror films give us the opportunity to feel terrified in safe surroundings. In this way, through the game of humor, you can test where these moral boundaries lie. The punitive reactions will only come later, if you cross more dangerous boundaries.

"Humor and tickling are very similar, because humor enters a person's intimate zone: you have to get permission to do it," Johansen says. "You can't tickle someone who doesn't want to be tickled, because they won't laugh. You can say something that crosses the line, but the other person has to recognize that this is what's happening, and in that way it resembles touching. It's a little scary and slightly illegal, and it happens *between people*."

In the past there was the carnival, where slaves became masters and masters became slaves, and where humor and transgressions went hand in hand. The reason for having a carnival back then was that, until the Renaissance in the 1500s, people lived in almost absolute hierarchies. To endure the extreme differences, the rules had to be broken now and then; if the slaves were allowed to play masters for just one day, they might not want to have a revolution. It was a kind of safety valve. Now we live in an age where social mobility is possible—or seemingly, at least—and Western society is no longer as feudal and rigid as it was. We live with constant change, and the boundaries of human behavior are continually shifting. Nevertheless—and perhaps precisely because of this constant change—we need humor more than ever. The comedian is like a modern-day exorcist, who now collects cultural tensions and ambiguities and purges them from our consciousness—it's not strange that the protagonist in Michel Houellebecq's novel *The Possibility of an Island* is a disillusioned, middle-aged comic. There are many rules and boundaries that are in constant motion and that we have to understand. And to deal with them and build a sense of community, we have to laugh at them too. We need to have nonsense and joy.

"Having no sense of humor is like having no sexuality; it means missing out on some of the best things in life," Johansen says.

For him, humor is mainly about participating in community, testing the limits of community, and, at worst, risking ostracism and complete loneliness.

"This is what separates humans from all other primates: we have a more sophisticated way of dealing with violations of the norm that involves many nuances, one of which is humor," he says. "If you laugh yourself when a group is laughing at you, you will remain part of the community because you are seeing yourself as part of a bigger situation—from the outside. You are being self-deprecating."

Being self-deprecating can save you in a group: it shows that you are seeing yourself through the eyes of the community. But grief can often blind you to the absurd. And when you are grieving, it's easy to feel that nobody understands you; everything becomes difficult, your inner space is dark and closed off. Your entire world is permeated by sorrow. This often makes you unable to laugh with others and puts you outside the community. Depression and anxiety make you less humorous. Death should represent the end of humor, and of community itself. But community is nevertheless always there, and in many forms, even if it is hard to see.

Therese Tungen lost her son, Edvin, when he suddenly died of a brain hemorrhage at only six years of age. When the day of Edvin's funeral came, it created a moment when Therese felt less lonely: everyone was preoccupied with her son. The worst thing for her is when no one talks about him, when they pass over what happened, in silence, and he becomes invisible, because for those who loved him and miss him, he is always there. But after the funeral, the impossible happened. That evening, her husband and their daughter were watching YouTube videos when he began laughing at a funny video clip. "Bår laughs with his mouth open. It had been a joyless day, but even then, there can be laughter. He points ahead," she writes in her memoir *Snu deg: Edvin's bok* [Turn round: Edvin's book]. Laughter implies a kind of hope.

Else Kåss Furuseth created two comedy shows based on her own grief, after both her mother and her brother took their own lives. From the stage, she has talked about suicide and the time that follows, about how to go on living after something so terrible, and about mental illness.

"Is it possible to laugh at something that's almost too difficult to talk about? Not because suicides are funny, but because they *are* so incomprehensibly sad. And when something is that sad, I'm extra happy that at least something is still fun occasionally," the comedian says of the idea for her shows *Condolences* and *Congratulations*, which were both staged at the National Theater in Oslo. "The dream was to make the audience laugh, genuinely, not just in a tragicomic and compassionate way, but because we are still alive. Humor keeps me going."

So humor is about holding on to life and each other. Grief is one of the loneliest things you can experience, while humor is a form of medicine.

"It's such a relief when the audience laughs and I know they've understood me! The opposite of that—that they didn't understand—would be a crisis! There's so much at stake: if they don't understand, it's stupid for them to have bought a ticket and come along. It also makes it a very lonely experience for me," she says of standing onstage and risking something, because a failed attempt at humor can go very wrong. If she's onstage and hears silence when there should have been laughter, it leaves her feeling totally alone and abandoned. So all jokes are part of a high-risk game, some more than others. The closer you get to the limit of what is acceptable, the more you have to lose.

"What I like most is transgressive humor, crossing boundaries, and daring to joke about things that you shouldn't joke about," says Kåss Furuseth.

This means she is risking the great silence, the condemnation, the scandal. But she is clearly willing to take that risk, because joking about difficult themes is too important not to do.

"When my brother died, I found out that he'd been planning to watch the nine-hour-long Holocaust documentary *Shoah* that week," she says. "And I said to my dad, 'If I'd seen that, I would have killed myself too,' and he laughed and said, 'But you mustn't say that to anyone else, because they wouldn't understand,' and then I thought, 'But I want them to understand, so I'm going to create a show about it.'"

Her brother's suicide had triggered something that might seem impossible: a joke. She then collected other taboo jokes, about suicide, about her family, and began piecing them all together: like the time a journalist asked if she had considered suicide herself and her thoughts on Norway's most tempting suicide locations. (Some towns are uglier than others.)

"So the humor is about connecting with other people?" I ask.

"Yes, it's about being together, about community."

Kåss Furuseth first realized how very much she needed humor when her mother took her own life.

"Some of the girls in my class just called me 'Humor' because I goofed around so much, but when my mother died they stopped because they didn't want to hurt me," says the comedian. "And I remember how sad that made me feel, to be told that I no longer needed to be funny."

But can laughter unite an entire community that has been terrorized, when thousands are suffering shock, grief, and PTSD? A terrorist act has nothing to do with laughter, and very few traumatized people manage to laugh together. We humans become crippled by fear when something as dreadful as terrorism occurs; we become timid and closed off. I remember the July 22 attacks in Oslo, and the grave atmosphere that hung over the city after the bomb, all the flowers piled up in front of the cathedral. People were nervous and vulnerable and wept openly. We were anxious and skittish, like Toffle. But a few months later we were able to laugh, together. In Room 250 of Oslo's main courthouse, the killer, Anders Behring Breivik, sat behind bulletproof glass, totally isolated and expressionless, while the survivors of the Utøya massacre came forward, one by one, and recounted how he had chased them around the island, how they had seen their best friends die, how they had played dead to avoid being shot themselves. For ten weeks, the entire country almost ground to a halt while the worst crime during Norwegian peacetime was examined in detail. Journalist Stian Bromark was following the trial from the public seating area when one of the survivors, Ina Libak, made the whole courtroom laugh. Libak had been shot four times, and yet she still managed to turn the painful courtroom situation into one that forged community through humor.

"The contrast between what she said and how she said it couldn't have been any stronger—she made us laugh and smile despite the pitch-black circumstances," writes Bromark. "It was the restrained way that she showed her scars to the court, before quipping that the facial scarring was hard to see because she'd spent so long covering it with makeup.

"That day we learned something important about our indomitable will to live, about human compassion in practice and about how optimism is contagious," he continues. "We laughed out loud because at that moment, in that place, it was the very best way to grieve."

Laughter as a release can move us out of a state of deadlock and into a community. A shared joke can help us view people and situations differently and see ourselves from the outside—it offers a new perspective that makes our own situation easier to understand, maybe, or helps us find light where there seems to be nothing but darkness.

But anything involving a strong community will have a downside. One day my daughter revealed to me that her classmates had been laughing at her best friend. The little boy had been very upset and had begun lashing out, trying in vain to fend off the bursts of laughter, which was of course impossible, like fighting smoke, and it just made the children laugh at him even more.

"But I didn't laugh at him, Mom, I really didn't!" she said.

As perfect as laughter is for creating a deep sense of community, it is also perfect for creating a strong sense of exclusion. Laughter can also be a doorway to loneliness. Few communities include absolutely everyone. The Greek philosopher Aristotle believed that certain types of humor should be forbidden, because they were malicious: "Most people enjoy amusement and mockery more than they should . . . for mockery is a species of revilement, and legislators prohibit certain kinds of revilement, and perhaps they should have prohibited certain kinds of mockery also," he writes in the *Nicomachean Ethics*. Humor can, as Aristotle claims, possess an element of ridicule—and that is pure ostracism, as my daughter so painfully experienced that day at school, when everyone laughed at her best friend.

"We tend to overestimate how much kindness there is in humor, because it also contains aggression," says Markus Johansen. "Remember that anything that creates unity can seem correspondingly exclusionary, and humor is one of the strongest social bonds there is. There are many contradictions to how it can be used. With humor, we have assembled a highly complex social tool."

Nothing is lonelier than being the subject of malicious laughter. If a whole group of people is pointing at you and laughing, you will, of course, feel lonely; what you are experiencing is clear ostracism from your community, which is almost impossible to ignore or to fight, as my daughter's schoolmate found as he punched the air to fend off the volleys of laughter. Even if a few of your classmates or colleagues are just laughing hysterically without you knowing why, without it even being aimed at you directly, you'll probably feel very left out. Humor that creates a safe community for some can be a dark nightmare for those who do not belong.

For Hilde Susan Jægtnes, school was a kind of battlefield. She was ridiculed for the way she spoke, for the way her body looked, for who she was, and for her extracurricular interests, which were classical music and horseback riding. She did well at school and liked goofing around and drawing on the blackboard, but all these things turned out to be "wrong" and became material for the bullies' cruel jokes.

"I wasn't cool, not in the slightest," says Jægtnes today.

As an adult, Jægtnes moved to England and then to the US to train as an actress, musician, and screenwriter. In London, she attended Rose Bruford College, then earned a master's degree in screenwriting at the University of Southern California in Los Angeles. She has written scripts for a number of feature films and television series, has published poetry, and is an award-winning and critically acclaimed author. Her quirky sense of humor and total disdain for hierarchies and popularity contests, all previously sources of her outsiderness, followed her into adulthood and are now considered her strengths—the distinctive qualities she refused to smooth out while attending secondary school.

"During the years I was being bullied, I refused to change my taste in music, for example, just because I wasn't listening to the 'right' kind of music."

At home, her parents were kind and understanding, and when she began secondary school she had a few friends, although they were in a different class. But on her first day at school, she met a funny and intelligent boy and developed a crush on him, but he would turn out to be her biggest tormentor. He was in the same class as her—and knew how to use humor as a weapon.

"People are naive about what children can do out of mischief," Jægtnes says.

The main bully had the entire school in the palm of his hand. He would make teachers break down and cry in the middle of a lesson. No one stopped him, not even the principal. Jægtnes knew that telling her parents about the bullying wouldn't help. Her kind and supportive mother would tell her every day that she loved her, but it didn't help teenage Hilde much when she left the house.

"I felt like a failure, like I was worthless," she says. "I had panic attacks and found it more tempting to stay at home. I would say I was sick with asthma to avoid going to school. Then sit at home eating chips and reading." New research shows that the consequences of serious bullying can be as damaging for children as domestic violence. The resulting traumas are equally significant.

"I was very lonely," Jægtnes recalls. "But I never talked about it. I didn't want my status to decline further, so I pretended it didn't bother me. At the same time, I fantasized about having a remote control that when pressed and pointed at my bully, would turn him into a skeleton."

She instead found an entirely different kind of friendship— with animals. Being around cats and horses involved having a deep sense of acceptance and belonging from living creatures that had neither language nor a sense of humor. She had cats and a parakeet and took up horseback riding. The cats followed her everywhere, climbing trees with her and curling up beside her as she read. And after school she would spend hours in the stable, her arms wrapped around the horse, stroking its soft coat. She dreamed of moving into a stable and becoming a vet or a riding instructor.

"I think I loved animals more than people at the time," says Jægtnes. "A relationship with animals is far less complicated—it is totally unconditional love. The contact with the big horses is very physical; you only communicate through touch, it's such a clear language."

Despite all her successes as an adult, despite having appreciative friends and colleagues, a loving husband, and an exciting job, the bullying has affected her and continues to affect her.

"I can be suspicious of people," she says. "I have difficulty trusting people. I've become very skeptical. I'm probably very social, and I like people, but I've got a system for testing the people I meet and do a lot of analyzing to find out if they're trouble or not."

Seeing a gang of young teenage boys will make her pulse quicken from anxiety, despite her now being an adult and too old to feel threatened by secondary school boys. But these gangs can still remind her of the helplessness she once felt and prompt her to consider ways of escaping the potentially "dangerous" situation.

The type of bullying Jægtnes suffered made her feel as though she was in a warlike situation. She had to be constantly on guard, a state psychologists call "hypervigilance," in which you behave as though any new situation is potentially threatening. You become unfocused, restless, and afraid; you sleep badly. And it's no wonder, when you are constantly afraid of other people and an attack can come when you least expect it, from the people around you—those you should really consider to be a safe flock.

Researchers at Keele University believe that there is a clear connection between bullying and humor. In 2013, the research group studied 1,234 children between the ages of eleven and thirteen and made a strange discovery: among the children who had a self-deprecating sense of humor—those who made fun of themselves for the other children's amusement—the likelihood of bullying increased. In other words, making yourself seem comical to make others like you will trigger their contempt.

"What our study shows is that humor clearly plays an important role in how children interact with one another and that children who use humor to make fun of themselves are at more risk of being bullied," said the lead author of the study, Claire Fox, without reflecting more on whether it can sometimes be the opposite: that bullying will trigger the self-deprecation, as a pure survival mechanism.

Bullying is prevalent in Norwegian schools, and approximately sixty thousand children suffer bullying regularly. Another growing trend is cyberbullying. When 26,652 of the country's pupils responded to a survey, the results showed that 5.8 percent of Norwegian schoolchildren are bullied, either online, physically, or through bad-mouthing and exclusion. Girls experience bad-mouthing and exclusion far more than boys, while boys experience more physical bullying. And although society seems to have become less brutal than it once was, the level of stress appears to have increased among schoolchildren.

In the last ten years, bullying and violence between children has risen by 75 percent. The research also shows that those who experience violence at home are more likely to become victims of bullying at school.

"More children are registered as victims of physical violence, threats, and other forms of ruthless behavior now than ten years ago," says Reid Stene of the Norwegian statistics bureau SSB.

Excluding someone through humor isn't just reserved for children either. A Norwegian-Danish research group recently studied the use of humor in the military, to see if it was somehow connected to the large drop-out rate of female recruits. They set out to search for a direct connection between exclusionary humor and actual exclusion, and found that the complex role of humor in the military is to help these institutionalized young people bond as they live under quite stressful conditions. But this humor occurs entirely on the men's terms and is a very sexualized type of humor that the female recruits are naturally unable to participate in.

"Being the social creatures that we are, we'd rather not spoil the good atmosphere, because that would mean no longer being part of the group," the research group points out. "We tend to join in and laugh, even when we perhaps don't find it funny."

So this harassing humor is never reported. What female recruits do instead is quietly disappear from the services. "But the best reaction to this harassment is when those observing it no longer watch and laugh, but instead react negatively to the offending colleague's behavior," the researchers note. This means that changing a culture of humor within established institutions like the police, fire service, and military can change the composition of genders. But it is difficult. Nobody wants to interrupt a joke. At first glance, a funny comment may seem quite harmless. Bullying, however, is very dangerous, and bullying and humor are often connected. Among the long-term effects of bullying, researchers at Leiden University find a heightened risk of suicidal ideation and suicide. People who are bullied are six times more likely to have suicidal thoughts than the control group. And this is a pattern that repeats itself in the workplace: in Norway alone, more than 250,000 people experience harassment, ridicule, threats, violence, and bullying at work annually, according to a survey by the country's National Institute of Occupational Health.

And the perpetrators of this bullying are affected by it as well. Or are they affected before they even begin? What's striking is that people who were bullies at school end up with lower-than-average education, have greater problems at work, are three times as likely to be unemployed, and are almost twice as likely to be at risk of substance abuse. Bullying others is a lonely act; you don't make real friends by ridiculing others. It's also possible that many bully others to avoid being bullied themselves or as a result of difficult circumstances at home. This became clear when 1,266 adults were studied by a research group at the Regional Centre for Child and Youth Mental Health and Child Welfare, Central Norway—years after the participants had finished school. The results show that many who bully have also experienced bullying themselves. In a large survey carried out by the British anti-bullying organization Ditch the Label, which involved over 7,000 young people, 1,200 of them said they were active bullies. Many of those respondents came from low-income and poorly educated families. And a large percentage of them had endured traumatic or turbulent experiences in the last five years that may have been stressful; they also had low self-esteem. The figures also showed that those who had experienced bullying were twice as likely to bully others. It was a fairly repetitive pattern of violence. A bully can be skilled at formulating a cruel joke but not necessarily good at the other, more vulnerable parts of a friendship.

This was probably the case for Hilde Susan Jægtnes's bully, who, among other things, ridiculed her for her regional way of saying the letter R, although the strange irony was that he used exactly the same throaty pronunciation. As an adult, he told Jægtnes about his own background, and yes, he, too, had been bullied. Jægtnes, however, doesn't feel sorry for him, because what she suffered didn't turn *her* into a bully. On the contrary, she became someone who tries to help other people; if someone at a party seems a bit lost, she'll approach them, and when she sees bullying and violence, she'll try to stop it. She doesn't think he deserves her pity.

"I've almost forgiven him now. But only almost," she says.

In one way or another, Jægtnes's bully has lost his power over her, though the scars of the bullying never really go away.

I ask if the fact that she was bullied affected her sense of humor.

"No, I retained my sense of humor! In the period when I was bullied, I clowned about even more with the people I was close to," she says.

Because laughing feels so good, we can easily forget how powerful jokes can be as a means of exclusion. Cyberbullying is rapidly increasing, as is the use of humorous memes and online ridicule. Forty percent of those who report digital bullying are being bullied by someone they don't know, while more than 28 percent are being bullied by someone who doesn't attend their school but knows them. Some of this increase in digital bullying can therefore be attributed to bullying outside school. Cyberbullying doubles the risk of self-harm, suicidal ideation, and suicide attempts. Some researchers point to the connection between malicious humor and cyberbullying, although this is a field that has been surprisingly under-researched, given that Reddit, 4chan, and 8chan—and many other websites often associated with right-wing extremists—are driven by partly malicious and transgressive humor. For example, on 4chan there are racist jokes about lynchings and sexist jokes about subservient women.

In 2021, two researchers, commissioned by the European Union's Directorate for Migration and Home Affairs, presented a report about online right-wing extremism in which they pointed out how humor is used as part of radicalization: "Humour has become a central weapon of extremist movements to subvert open societies and to lower the threshold towards violence," they wrote.

By using funny and playful ways to communicate their ideology, extremists ensure that their memes and tweets are picked up by ordinary people and widely distributed. Members of the alt-right movement in the US have been experts at this, and the phenomenon is now spreading globally.

"This predominantly online movement set new standards to rebrand extremist positions in an ironic guise, blurring the lines between mischief and potentially radicalizing messaging," the researchers explain. "The result is a nihilistic form of humor that is directed against ethnic and sexual minorities and deemed to inspire violent fantasies—and eventually action."

There is a direct line from dark, transgressive humor to ostracism and pure violence. Anyone who experiences these dark elements can end up on a scary path toward loneliness; you'll know you're at risk of ostracism or violence if you find yourself being laughed at, ridiculed, or threatened through memes and jokes. Having said that, laughter can be one of the nicest, most inclusive, and most playful things we humans can share.

Either way, it's into this colorful and frightening world of humor that my child will be venturing. Understanding all the subtle codes that humor conceals is one thing; mastering them is another. I'll have to conquer my fear of her being bullied and mocked by her peers, of her being subjected to cold irony and sarcasm. She doesn't have a smartphone, not yet, but there are murmurs all around: the other kids at school are getting them, and at an increasingly young age. There are parental controls that can be used to limit access to adult content, and school staff can ask the children to put the phones away during lessons. Nevertheless, these prepubescent children, with their unfinished brains and chaotic emotional lives, who haven't yet learned how the physical world functions, are being connected to an entirely fictional world, where dark irony can be a pathway to right-wing extremism and violence. And the only thing I can offer to protect my daughter from all this is a secure feeling of attachment and the ability to laugh until she almost pees herself. Will that be enough?

When she laughs like that—so much that she holds her stomach—I imagine that it must be an infinitely powerful way of defending yourself from cruel humor. There are brief moments in my life when we laugh together and it feels like everything falls into place, as if the doors have opened wide and everything is perfect. It may be a flimsy barrier against loneliness, but it helps. On a bad day, it helps.

"Mom, did you just fart through your nose?" my daughter roars before collapsing from laughter, which in turn makes me start laughing.

"No, now it's bedtime," I finally say, trying to force my face into a grown-up expression.

I then, once again, read the story of the invisible child to her, more carefully this time. She curls up beside me, listening with her mouth half-open and an apple wedge in her hand, and suddenly it dawns on me. The invisible child is invisible for a very specific reason. She lived with an aunt who was "ironic all day long every day, and finally the kid started to turn pale and fade around the edges, and less and less was seen of her." The cause of Ninny's invisibility and loneliness was the ice-cold humor. But just as being invisible made her feel lonely, she feels connected when she finally becomes visible. And when she saves Moominmamma and becomes visible, she begins to laugh—a real, spontaneous laugh. She has become mischievous and lively. For the first time, Ninny has become a full-fledged part of a community.

"'Oh dear,' Ninny was shouting. 'Oh, how great! Oh, how funny!' The landing stage shook with her laughter."

4

Why it is so important to understand each other

IN AN ONLINE edition of the *New York Times*, there is a report featuring a short video of a man who is the last of his kind. He speaks the Taushiro language, and his name is Amadeo García García. The entire tribe to which he belonged, who lived isolated in the Peruvian part of the Amazon, has died out, disappearing one by one, until only Amadeo and his brother Juan remained. When Juan then died of dengue fever, the entire Taushiro tribe's culture and language rested on Amadeo's bony shoulders. It has now been years since Amadeo had a conversation in Taushiro. He still speaks it when he dreams, but to make himself understood when speaking to other people, he must speak Spanish.

"I feel sorry for Amadeo because he's a lonely man. You need to have someone to talk to," says Mario Tapuy, a resident of the village where Amadeo now lives.

Amadeo had to leave his abandoned village and move to a neighboring one, where he lives among people who don't speak his language and know nothing about Taushiro culture. He did get married and have children, but his wife left and the kids were placed in foster care because he became an alcoholic.

"We wanted him to become a teacher, and tell us about Taushiro culture and its language," Tapuy says. "But he was always drunk, so it was impossible. And now he's too old."

That it's possible to be lonely when you live in a small, tightly knit village may seem incomprehensible; most lonely people in cities can only dream of such an existence. You can't be lonely in a village—everyone knows each other! But this story is missing one important element: understanding. Nobody in Amadeo's new village really understands him.

I've been thinking about the word *understanding* increasingly often. It appears loneliness partly consists of a pressing need to be understood. My daughter regularly yells, "Mom, you don't understand!" when I'm pushing her to do something she doesn't want to do, like leaving for school on time. My husband doesn't understand me—nor *can* he immediately understand—when I cry suddenly because something has reminded me that my father is dead, and so in that moment I feel more alone in my grief. For his part, my husband is British and often feels lonely in Norway when Norwegians don't understand his background and references. But why is being understood so important? The landscape we are now entering is marshy and covered in a misty haze. We all want to be understood for who we are, but this can be extremely difficult. We are shrouded by lies and shame, detached from those around us by fear and grief, and this can make the bridges to our island break in the middle.

You feel it intuitively: you need to be understood by the people around you, and if you aren't, you are lonely and lost. What being properly understood actually means is hard to grasp, which makes it one of the boundary stones around the land of loneliness. But it's an issue I see alluded to in novels and in arguments in news columns; I've also found research showing that *not being understood* triggers pain, just as rejection does. It occurs in the same part of the brain and is experienced in a quite similar yet milder way.

In an experiment, Stanford University researchers used fMRI to scan the brains of people as they were being understood and misunderstood, documenting which parts of the brain were activated. The researchers believe that understanding is closely linked to the brain areas associated with attachment, while being misunderstood is associated with negative emotions. This isn't that surprising, and we perhaps don't need fMRI to tell us that. Nevertheless, when someone understands you, you bond with them. Those doing the understanding likewise bond with you, and as a result you become less lonely. But understanding someone is infinitely more complex than just reassuringly saying, "I understand." And the huge diversity of cultures has exacerbated this problem. In the Bible, the Tower of Babel reaches for the sky, built by people who, thanks to their unique ability to cooperate, succeed in erecting this overly tall structure. It is an expression of hubris, pure arrogance, a colossal project that God quickly destroys, scattering the people to the winds, each with their own language, making it impossible for them to understand each other. And while speaking English will soon be possible anywhere in the world, it is still difficult to be understood. Not only are there different cultures and customs, families and traditions, but also groups of people with infinitely varied memories, dreams, characteristics, and talents. We are the opposite of anteaters and pandas, who don't require much energy when trying to understand each other since they're pretty much the same. Everyone in their group has the same survival strategy. Anteaters live in one environment and eat one type of food: termites. Pandas exist similarly in China's bamboo forests. It's no coincidence that humans are the most successful species on Earth: we are so diverse and different. We have conquered all the world's climate zones, we eat a wide variety of foods (from caterpillars, snakes, and grasshoppers to cows, crabs, and whales), and we can do and want a huge variety of things. We are extroverted, introverted, esoteric, practical. So of course it's harder to understand each other.

But we absolutely must: it is the key to our survival as a species. And to ensure this survival, we are equipped with the very uncomfortable sense of loneliness that makes it imperative for us to be understood and drawn into the community. Understanding each other ensures that we can cooperate, which is so vitally important to our existence. Feeling understood, the research shows, is associated with attachment, joy, and good health, while not being understood is associated with the opposite. In other words, there is something fundamental about our desire to be understood; it is an antidote to loneliness. If we are to use the perhaps slightly vague definition of loneliness made in 1982 by the researchers Letitia Anne Peplau and Daniel Perlman—"Loneliness is not getting the social contact we desire"—this is probably what they mean. What you want isn't to meet as many people as possible, as you might do when encountering hordes of other travelers at a busy airport; you can do this for hours or days without feeling properly in contact with anyone. What you *want* when you meet other people is to be understood.

To learn more about what understanding actually *means*, I returned to the books I read when trying to understand the history of ideas as a student and found it to be a recurring problem: How can we understand someone from another era—when everything was so different? I can hardly claim to understand how life was for people living in rural Norway in the 1800s or in France during the Middle Ages! The early-twentieth-century author Sigrid Undset nevertheless confidently claimed the opposite: "Customs and traditions may change a lot, so much gets chiseled away throughout history, and people believe and think differently about many things. But people's hearts don't change at all." Undset, one of Norway's literary giants when it came to writing historical novels, was absolutely certain she could understand people she had never met, and from a time she had never experienced. And in many ways she was right, since our brains haven't changed structurally in the last three hundred thousand years. But this problem has appeared more often recently, during the culture wars and identity politics of today: How can people with different cultural backgrounds and political viewpoints understand each other?

To understand historical events and texts, claims the philosopher Hans-Georg Gadamer, you have to move in a kind of "circle of understanding." Your initial understanding of the unknown is based on preconceptions and context. After that, you examine things more closely, adjust your understanding, zoom out and examine the context again, return to the details, understand more of the context, which in turn sheds new light on the details, and so on and so forth in circles. And since you, too, are a participant in this, one that is in motion, dancing in a circle, your understanding of yourself and the situation expands into increasingly bigger circles. So you can never understand anything entirely—only more and more. It's perhaps similar to the breaks and microrepairs that attachment psychologists describe: the elaborate spiral that constitutes understanding and growing, together.

It's imperative that we always see ourselves in context, believes Gadamer, and how we are bound by culture and our shared humanity. "If we put ourselves in someone else's shoes, for example, then we will understand him—i.e., become aware of the otherness, the indissoluble individuality of the other person—precisely by putting *ourselves* in his position," he writes. Understanding is more than simply looking for similarities between yourself and other people, or projecting yourself onto someone else. Recognizing the individual differences, those that clearly *are* different, while having a basic understanding of shared humanity—*that* is understanding: "It always involves rising to a higher universality that overcomes not only our own particularity but also that of the other," Gadamer continues.

We are in fact *all* connected, by culture, language, and humanity, which all give us keys to understanding each other, despite our differences. Defining what understanding means is hard because understanding is a gradually unfolding process. We can never fully understand anything. But we still need to be understood, perhaps especially when we need care and protection, or in our close relationships. A baby, who is at the mercy of their parents, needs them to understand the difference between tired crying and hungry crying, and to realize that this crying is a plea for help, because if they don't, the newborn could be in danger. A child returning from school bruised after being bullied by her classmates needs her parents to understand this, so they can comfort and help their child deal with her feelings and perhaps resolve the situation. A woman grieving her dead child needs the people around her to understand, so she can recover within a protective circle of care.

Friendship is about being understood, and we want and need our family to understand us. Being understood makes us feel like a valued member of the community, that we have become part of someone else's inner world. That they are reflected in us and we in them enables us to feel their pain and them to feel ours. We look them in the eye; we laugh together at the same jokes. When we are part of each other's inner worlds, we take part in each other's stories and dreams and memories. Ideally, we also understand, with a form of respect and love, the things that make us especially different from each other. This is, of course, what John Cacioppo is talking about when he says that the human brain is social: the opposite of being lonely is being connected to another person's feelings and needs, without losing sight of our own. Being human is something that happens when we are together; it is not something we do alone. Talking about people as being separate from each other is as absurd as referring to the different neurological centers of the brain as being separate from each other. The brain only functions when millions of cells join forces in large networks, in clouds of electrical activity. People only function if they are connected (more or less) to other people. At first, when we are babies, we are totally, and almost continuously, dependent on being connected to a caregiver. The importance of this connection becomes less crucial over time. But a person who never feels understood will be a lonely person.

The brain is a storyteller more than anything. Our experiences in life are colored by our expectations of how our life should be—by our own memories, feelings, and opinions—which makes it especially important to share our inner world with other people, so we can all take part in the same project. But while it may be colorful and distinct, and a source of ideas and happiness when shared, the uniqueness of our inner world also means a great deal can go wrong when meeting other people.

I can't be sure, but I imagine a conversation between two pandas could go something like this:

"Bamboo?"

"*Yes!* Bamboo!"

While a conversation between two humans could, for example, go like this:

"What would you do if you were me?"

"What would I do? Let's see, now... A noblewoman... who's fallen... I don't know... Yes, I do know."

[*Picks up the razor and makes a gesture.*] "This, you mean?"

"Yes, but I'd never do it, mind. That's how we differ."

"Because you're a man and I'm a woman? Does that make a difference?"

"It's the difference between men and women."

[*With the razor in her hand.*] "I want to, but I can't... My father couldn't, either, that time when he should have."

Granted, August Strindberg was overly preoccupied with conflict and how badly women and men understood each other, as demonstrated in his play *Miss Julie*, but still: you probably understand where I'm going. There are endless possibilities for misunderstandings and distance between us humans in a conversation; there are cultural differences and ideas about gender, there are anger and historical reference points, contempt and memories, repressed love and old dreams, and all these complex memories and feelings color our conversations and make it hard to understand each other.

To understand each other, we perhaps need to begin with something simple. Something like the pandas' conversation about bamboo. According to the Norwegian health authority's guidance for lonely people, one way out of loneliness begins with this superficial understanding: "Reach out to social meeting places such as groups that run organized leisure activities or those that need voluntary workers. If you have an interest or hobby, try nurturing it as part of a community. Try creating routines where you repeatedly meet the same people. This is how strangers can eventually become friends."

One of the more successful attempts to cure increasing loneliness among older men has been the so-called Men's Shed, which involves men getting together and fixing things. These places offer men somewhere to use their carpentry or metalwork skills, and when attended by men of the same age, with the same interests, the activity eventually results in friendships. The project began in Australia in 1998 and is now a global program with hundreds of projects in the UK, Canada, Ireland, Denmark, Australia, and the US. The understanding at a Men's Shed begins with something practical ("bamboo"), and can evolve into something more complex, because we usually dream of more than just saying "Bamboo" or "Pass me a hammer" to each other. Understanding is about opening your inner world to other people. The deepest understanding is almost mythological and is described in pop-song lyrics and poetry and by Freud when he writes about oceanic feeling. It is the feeling of two people almost merging. In his novel *I'm Thinking of Ending Things,* the author Iain Reid writes about how he imagines such absolute understanding:

I think what I want is for someone to know me. Really know me. Know me better than anyone else and maybe even me. Isn't that why we commit to another? It's not for sex. If it were for sex, we wouldn't marry one person. We'd just keep finding new partners. We commit for many reasons, I know, but the more I think about it, the more I think long-term relationships are for getting to know someone. I want someone to know me, really know me, almost like that person could get into my head.

This is deep understanding—brains in total unison, with activated mirror neurons, eye contact, and genuine empathy, a sense of absolute connection while acknowledging the other person's differences. This way of being with another person involves sharing stories and agreeing on what life is about and what goals you share, what you can contribute, and what the other person wants and does not want to do. It means not lying to each other, not hiding important things about yourself, but living in complete openness with each other. And just as loneliness is a health risk and causes early death, strong connections and good understanding are health-promoting. Loneliness research shows that being part of a religious group will be great for your sense of community, for how safe and well you sleep, and can even prolong your life: you share your dreams and reality with everyone around you, perfectly attuned to your surroundings. In the world's so-called blue zones, areas with a high proportion of people who live to be a hundred years old, you find people living in strong communities; in the US's only blue zone, Loma Linda, California, more than 30 percent of the population consists of Seventh-day Adventists.

But you don't have to join a religion to be understood. It is the specific way we understand each other that makes *talking* such a central part of modern therapy. This is what made Sigmund Freud and his theory of psychoanalysis so revolutionary: that talking, and sharing your whole inner world—sexual desires, regular nightmares, and so on—with someone nonjudgmental, could be forever life-changing. The irony is that Freud's psychoanalysis became so fashionable among the modern, urban bourgeoisie, where loneliness seems to thrive best behind closed-off facades. But the more impenetrable the masks, the more we perhaps need secret places where we can reveal our inner world, because there's nowhere else to do it.

Researchers are concerned that the numbers of lonely people are increasing most in densely populated areas: the epicenters of the loneliness pandemic are cities. Author Noreena Hertz believes that turbo-capitalism and modern architecture constitute an attack on our sense of community—cities that are constructed without public spaces and recreational areas, cities built to ensure quick profits for the richest, which provide few opportunities for random conversation or chance encounters that could lead to deeper friendships. There are benches that are painful to sit on, fewer parks, and balconies taking over from shared backyards, all of which make us rush through a city instead of using its urban spaces to meet other people. Neoliberal capitalism, especially the forces driving social media, which is specifically designed to hold our attention, has all but ensured that we don't meet each other properly at all. Our eyes are constantly drawn to a screen, and whatever is happening on social media, rather than to other people's faces.

Efficiency creates distance. It takes time and a little fumbling to understand a new person, to get a true sense of who they are. "And that went for all the people in the wood," sighs Moominpappa, when attempting to make his first friend. "The birds, the worms, the tree-spirits, and the mousewives seemed to be in a great hurry about something or other. Nobody wanted to look at my house, or hear about my escape, or about the strange way I came into the world," writes Moominpappa in his autobiography, penned by Tove Jansson. Urbanites may well identify with Moominpappa's frustration. The pace of the city, which ensures that factory owners and shop owners and internet entrepreneurs make lots of money fast, is at the same time a threat to unscheduled friendliness.

But a slow-paced village doesn't necessarily shield you from loneliness either, as the story of Amadeo demonstrated. His story is almost an echo of another, about the world's loneliest whale, who scientists call simply "52."

This fin whale—unlike the other members of his species, who sing their low-frequency songs at 20 hertz—makes sounds as high as 52 hertz. So he cannot be understood by other whales, no matter how much he cries out into the vast expanse of the ocean. 52 was discovered in 1992 and has fascinated scientists through the thirty years they have known him; he is even the star of a 2021 documentary, *The Loneliest Whale: The Search for 52*. The extensive studies of this creature show just how concerned we are with loneliness as a phenomenon and how fundamentally we understand it. Loneliness is about having no flock, and it can be as simple and as difficult as not being understood. Could it be that our cities are full of people metaphorically singing in different frequencies too?

While it's hard to define what being understood is, and even hard to pinpoint when it's happening, it's at least easy to know when it's *not* happening. If you don't feel understood, your inner and outer worlds find themselves in stark contrast to each other. When author Hilde K. Kvalvaag's teenage son took his own life, the people around her became intolerable. None of them were able to underszatand her—and she was unable to be part of their reality.

"My God, you are dead, and she's talking about furniture," she writes in the novel *Djuphavsslettene* [The abyssal plains], which is about her son's death. "I've lost a person that I love, and she's talking about furniture, it almost makes me angry. I hear her mention your friends, and stand up, almost shouting. *I can hear everything you're saying*, I say. *Could you please not say that.* 'It's good that you tell me,' says S. 'It's a good sign.' I'm sorry. I almost cried, but I can't stand hearing about friends who are living, it's too painful."

Kvalvaag describes what it's like to feel excluded from all social situations, how even the most innocent conversations are upsetting.

This could be what W. H. Auden's poem "Funeral Blues" is really about—about wanting to make the entire world understand the pain the poet is feeling. He describes a world of silence and stopped clocks, noiseless dogs and pianos, concluding with "Let aeroplanes circle moaning overhead / Scribbling on the sky the message 'He is Dead.'"

Only when the grief brings everything to a halt, indeed when time itself has ceased, will the poet's inner and outer world be in harmony. Only then can other people understand.

Our need to fit in with those around us means we are willing to make a lot of compromises. Appearing too unusual can be off-putting, so it will often feel necessary to hide yourself and your quirks for fear of other people's condemnation. You won't share the things that matter, and thus you deny people the opportunity to understand you. An important driver of such secrecy and lying is shame. But the paradox here is that we feel shame and hide ourselves mostly to gain other people's acceptance. We tell lies and keep secrets in order to appear "normal" and perhaps more desirable, but the price we pay for this is never feeling understood.

"Loneliness and shame go hand in hand, because shame creates loneliness. But loneliness is at the same time an example of something very shameful, at the very core of what we are ashamed of. So there is a self-reinforcing power in the loneliness and the shame. You become lonelier from the shame, and more ashamed, and that makes you even lonelier," says Helene Flood Aakvaag, who has a PhD in shame and trauma and works as a researcher at NKVTS, the Norwegian Center for Violence and Traumatic Stress Studies.

"This whole spiral is compounded by the fact that loneliness can be interpreted as something that is wrong with you and is therefore the reason people don't want anything to do with you," she says. "They act like it's contagious. They withdraw. And you become lonelier."

She has seen that those who experience violence become, unsurprisingly, more withdrawn and lonelier, with shame and the fear of other people's judgment being driving forces. And those who suffer violence are right about this, unfortunately. Research on victims of violence and rape shows, incredibly, that victims who share their stories often find themselves accused of being the cause of the abuse they were subjected to. As often as they are cared for and understood, rape victims face suspicion and disbelief.

"There are many reasons why people withdraw from victims of violence, one being that the violence also makes those close to the victim feel unsafe," Aakvaag observes. "They worry that they too could be affected. The fact that the violence is occurring at all can threaten their worldview. It can then feel important for them to ask if there is something about the victim that may have caused the violence. Doing so can attribute some kind of logic to the violence, which can in turn maintain your own sense of safety. It's not real safety, of course, and this way of dealing with the violence has a great cost for the victim, who in addition to the violence now feels great shame. Because it's their fault that the violence happened."

When we blame victims of the violence they suffer, we push them out of the community as if they are some kind of defective product—we give them a sense of there being a fundamental problem with them, one that makes them unworthy of being understood or looked after by their flock. That is loneliness.

It is this kind of shame that has affected the children and young adults who were shot at while camping on the island of Utøya during the July 22 terrorist attacks in 2011. Many believed that the unarmed people on the island should have attacked the heavily armed man, who was dressed as a policeman, and that the adults on the island had been cowards. One in three Utøya survivors has received hate messages. Eskil Pedersen, the then leader of the Workers' Youth League, which was having its annual summer camp on the island that day, was especially targeted. Pedersen was called a "traitor" and along with many other survivors received hate messages and death threats. Some people even urged the attorney general to sentence Pedersen to six years in prison. In the subsequent five years, he received thousands of threats and hate messages, the first of which came as early as the night after the attack, while he and the other victims were sheltering at a nearby hotel after being fired upon for over an hour. At a time when they should have been understood and protected, many were being subjected to a campaign of pure hatred and exclusion. This behavior is commonly referred to as *victim-blaming* and inevitably causes loneliness.

There was, however, someone on Utøya who did think he understood—one man who, for a moment during the massacre, felt connected: the terrorist, Anders Behring Breivik. Instead of firing the rifle he at one point aimed at a blond-haired young man called Adrian Pracon, Breivik lowered his weapon. Pracon in fact represents everything the terrorist was out to destroy—something Breivik had described meticulously in his 1,518-page cut-and-paste manifesto. Breivik firmly opposed inclusion, community and integration, tolerance and diversity, all values that Pracon, the son of Polish-Catholic immigrants, believes in and has been willing to sacrifice a great deal for. Nevertheless, the terrorist thought that just by looking at Pracon, they understood each other; he believed that he saw himself. And that is why Adrian Pracon is alive today. Yes, he and the terrorist have one thing in common: they want to be understood. But while Anders Behring Breivik communicated his ideas about his worldview by hammering out a manifesto and sending it by email to 1,003 people, Adrian Pracon has been concerned with being understood in a totally different way.

Adrian has always been very boyish: he played sports and was part of a large and active male group. At one time he would go hunting with his father and has passed the hunting test. More recently, he also learned to ride a motorbike.

But although everyone felt they knew him, that they understood him, there was something he always had to keep hidden. He didn't consciously live in shame or fear, but for a long time pushed part of himself away, a part he didn't even understand himself and couldn't face dealing with. It was such an uncomfortable secret that he even kept it from himself.

"I constantly felt like there was something wrong with me," he says.

His parents had traditional jobs; Mom was a nurse and Dad worked in a factory. There was no reason to think that Adrian, their youngest son, wouldn't get a secure job and a wife when the time was right. Polish Catholicism is conservative and forbids divorce and abortion. That Poland was the birthplace of Pope John Paul II is part of Polish-Catholic pride, and the battle against so-called moral decay has been growing steadily there in recent years, along with right-wing populism; there are now "gay-free" cities in Poland, where queer people are openly discriminated against. Any boy growing up in this environment will have to repress part of himself should he develop feelings for other boys. Feelings like these can't be there.

"I decided that this wasn't who I was," Pracon says. "I joked for years about gays and used the word as a slur. Because it wasn't me, it was something else, I denied that I felt attracted to men. I hated that I did! I was terribly conflicted. That word, *gay*, it had nothing to do with my feelings."

Nevertheless, rumors persisted. He wasn't like the other guys, not really. Shame and the fear of being rejected or punished make us withhold information, hide things and cover up, distort the truth. And painful secrets like these make us lonely, because if we don't share them, we can never be fully understood. This is precisely what Pracon experienced. By the time he was sixteen, in the summer before his second year at high school, it was obvious something was going on and that his family didn't know everything about him. Something important had to come out.

"There was no way around it. I had to say something," says Pracon. He had been in a relationship with a boy. The rumors were entirely true.

When his mother heard that her son was gay, she called the priest. She struggled to accept that her son wasn't what she

93

had thought he was. So although he had started understanding himself, he was not understood by his family. And his sexual orientation was met with fear.

"There was a lot of talk about how I could die of AIDS, and there was also talk about how I might grow out of it—the sexual orientation, that is," he says.

So when Pracon told his family that he was falling in love with boys, he cut ties with them the same day and moved away from home. And this had major consequences for him: after separating from his family, he dropped out of high school, because no longer living at home meant he had to get a job to buy food and pay the rent. But he also made gay friends and became part of the local queer community, which organized parties and events in his hometown of Skien.

"There was quite a lot of partying, which became quite superficial after a while," Pracon says. "It was social and fun, but I needed some everyday friends. I needed everyday life and a bigger project. For a while I considered working with animals, because animals can't tell lies like we humans can. I felt let down and misunderstood by my family and friends."

Being young and gay in a small Norwegian town can often involve a heavy burden of shame. Shame is unfortunately a big part of queer life, and there are many people with coming-out stories like Adrian's. Some are far worse.

Gay history is full of heartbreaking stories about women living together as "sisters" or "friends" to conceal their emotional lives, and men living out their love in secret—knowing that declaring it would lead to a harsh punishment. To this day, homosexuality is deemed a punishable offense in sixty-five jurisdictions, and in twelve of these, loving a person of the same gender can result in the death penalty.

Until 1972, under Section 213 of the Norwegian Penal Code, it was illegal for men to have sex with men—lesbians weren't even covered by the law. And although Tove Jansson lived with a woman—the artist Tuulikki Pietilä—for almost fifty years, the relationship was shrouded in secrecy: in 1985, when Tordis Ørjasæter wrote *Møte med Tove Jansson* [A meeting with Tove Jansson], the only authorized biography about the Moomins author, Tuulikki is consistently referred to as Jansson's "friend," not her girlfriend or partner, which was, of course, what she was. Tuulikki is the inspiration for Too-Ticky, a character who lives in a house by the water and who brings the invisible child to visit the Moomins. Without Too-Ticky, the invisible child would have remained invisible, and the story itself perhaps represents the love Jansson herself had found: that love had made her visible and colorful.

If you know that you won't be accepted for who you are— and that you might even wind up in jail for it—you have to hide. And the fear of not being understood lives on, even in Norway, where homosexuality was decriminalized fifty years ago. In Norway today, between one hundred thousand and five hundred thousand people can be defined as LGBTIQ+, but it's an uncertain figure, possibly because of the shame many feel—often toward themselves—that makes them disinclined to even tick an anonymous form.

"We in the queer community live with constant narratives about what we'll be like as adults, with partners and children, but they're things we don't fit in with. For my part, I was terrified of disappointing my family, or of not belonging in the same way anymore," says Ingvild Endestad, who for five years was head of FRI—the Norwegian Organization for Sexual and Gender Diversity.

"It can be very lonely," she says. "You never quite fit in. You don't know if your family, friends, school, or work will accept you. It's tough being a minority in your own family. Many people are only used to hearing derogatory words about being queer, only used to it being a dirty word. This forces you to describe yourself as perverted when talking about yourself, and to refer to your own love life in negative terms."

It took a long time for Endestad to come out, and when she finally did, she was afraid, despite the fact that her family had never spoken negatively about gay people.

The statistics for young people in the queer community are bleak. Members of this group are among the most likely to attempt suicide. Young gay and bisexual men are three times more likely to attempt suicide than the rest of population. Trans people are the most likely of all. Feeling both excluded and targeted of course makes you vulnerable to depression. Even in egalitarian Norway, it still isn't safe to openly live as gay; many experience threats and violence. In a 2022 survey conducted by the Norwegian Police Service, far more bisexuals, lesbians, and gays said they were victims of hate crime than heterosexual people. Forty percent of bisexuals, gays, and lesbians said they had experienced hate crime. Twenty percent of the population state that they find gays kissing or hugging disgusting. All the jibing and exclusion, be it subtle or overt, makes the person experiencing it feel lonely. To feel constantly different and not understood leads to a sense of estrangement from the community, which is again the very definition of loneliness.

On June 25, 2022, the night before Oslo's annual Pride parade, a busy downtown bar called the London Pub suffered a deadly attack, and as a result the main parade, along with several others that had been planned around the country, was cancelled.

Whatever the motive was, the shooting did not surprise the queer community as much as it did the general population. "I'm one of those people who loves being queer—but sometimes I stop and ask myself: Do I really want to be visibly gay today? Why is openness something I'm constantly deciding on? It's because I know that it can potentially lead to violence," Nina Bahar wrote in the newspaper *Klassekampen* two days after the shooting. "In the early hours of Saturday we were under attack. But we've never really stopped being under attack. It makes me so afraid."

This is why being understood is so important: it means that you feel protected and loved as you are. This is something that many never experience. Endestad's family did try to understand her, and she didn't break contact with them, as others did.

"Being gay and belonging to several minorities can make you especially vulnerable," Endestad says. "Like being a minority in your own family, in mainstream society, and in the gay community, for example, or being gay and belonging to Norway's Lutheran Free Church congregation. When these people come out, not only do they risk losing the social aspect of being in the congregation, some of them also lose God. It can be an enormous loss, a double loneliness. And the gay community isn't that great at allowing them to find God again either. Having more affiliations with minorities, however, can also mean having more communities to identify with."

So while living in a sect might seem like a healthy alternative, given the statistics for life expectancy, it of course doesn't apply if you're gay and constantly fear being outed; a gay Adventist won't feel particularly safe and understood. Not only is telling people who you really are scary, it can also lead to reprisals and, in the worst case, violence and ostracism. We often have a reason for keeping a secret, and it's usually because we are afraid of how the people around us will react to who we really are. So we lie about everything, from our mistresses and our embezzling to our bodies and sexual orientation.

I, too, attempt to conceal as much about myself as possible; I want to appear exciting, beautiful, and wise, and no place facilitates that better than social media. I have lied about my background for as long as I can remember, probably because I felt ashamed. And it has made me feel very lonely. It has been an invisible wall in every relationship I've had, because no one has been able to know me, not properly.

Living with secrets is painful; it separates you from other people and makes you terrified of being exposed, of losing face. Secrets are often driven by the fear of being lonely, of being thrown out of the community. You become evasive and lie to protect yourself, perhaps to maintain the appearance of being *normal* (whatever that means), because being normal means you still belong to the pack. Lying shields you from being hurt, and it also has its benefits as a social mask. We cannot be brutally honest all the time. We lie to fit into the community, to smooth over conflicts, to avoid being abandoned.

In Latin, the word for *mask* is *persona*—which means we could say that having a personality is linked to being partially covered by a social mask. For most of us, the social mask isn't that different from who we perceive ourselves to be. But cover-ups and lies are a double-edged sword. If the lie becomes too big a part of your life, it can also help create an illusion about who you are, and in turn create a distance from the people around you. If you identify too strongly with the lie, you start to create what psychologists call a "false self," a shell of lies around the dark core of a personality. And this creates a great paradox:

you are hiding in order to be part of the herd, while simultaneously feeling cut off, because the people around you cannot understand.

Cathrine Finstad collects other people's secrets—and illustrates them. She began thirteen years ago, receiving them from anonymous donors, and now has more than 150,000 secrets hidden in her archives. And they keep pouring in. Finstad, a trained illustrator, got the idea after discovering a similar project in the US that used postcards, but she decided to use the internet to make everything more accessible: illustrated secrets end up on the website Norske Hemmeligheter and on the Instagram account @norskehemmeligheter.

"Most of the secrets I get are directly connected to shame," Finstad says.

The account is a kind of psychologist's office or confessional, where you can tell your secret without having to commit yourself or change a relationship with anyone. Your secret is released into the world, like a helium balloon that rises into the sky and disappears.

"I think people fear the consequences of telling their secret to someone they know," she says. "Or that it's too painful to reveal your true feelings when you've created an image or a facade. The secrets can be about jealousy, or the sender might feel like they're not performing well at work, or feel left out among their friends. And there's the hidden grief of the middle-aged man over losing his hair, or the woman who can't face sharing her breast cancer diagnosis with anyone except her partner. The secrets I get are extremely varied and come from multiple generations and people from all over the country. It's a chorus of voices, a project which I hope will lead to less shame and more openness."

Studies show that men and women lie equally often. When women lie, however, it's mostly to avoid hurting others, whereas men lie mostly to seem better than they really are; women want to be kind, while men want to look good. Both motives are governed by the same perception: that we risk losing our place in the community if we are honest. By lying, most of us want to bind ourselves to others or protect those around us from social pain. You don't tell a good friend that you're skipping his birthday party because you're feeling a bit tired; you make up a white lie about a sick grandmother. You don't tell a colleague what you *really* think of her new dress; you say warmly, "Don't you look nice!" But these are the secrets and lies that work as social glue. Many of the darkest secrets, however, are painful to carry and make the people carrying them lonely. These big secrets are often about the scariest things in life: their loves and fears.

The fear caused by the July 22 attacks caused a major change in the type of secrets Norwegians harbored.

"It was the clearest change that I've seen!" Finstad says. "For the first time, I began to get secrets from people worrying about terrorism in the streets or about being killed for their political views. Now I get secrets from Utøya survivors and bereaved relatives who, years later, just cannot move on. With so many years having passed, people seem to expect them to have put it all behind them, when they in fact wake up every morning still feeling traumatized by the attack. To them, it's a dark secret."

Many people felt different kinds of shame and loneliness after the terror attack on Utøya. Ingvild Endestad, who was once very active with the Workers' Youth League, knew exactly who was staying on the island when the reports started pouring in about shootings at the summer camp. She knew the people there very well. The organization being attacked that evening was one of her most important communities.

"For me, the Workers' Youth League was a very strong part of my identity. It was my whole life as a teenager, and when I left, it was a huge loss, a loss of identity. It probably sounds strange, but the most shameful loneliness I ever felt came to me after the Utøya attack," she says.

"Because all my friends were there, and I wasn't part of it. And while I'm eternally grateful that I wasn't there, to be on the outside, watching the people I loved so much being so grief-stricken, felt lonely. But how could I even think such a thing? How could it feel wrong that I wasn't on Utøya that day? When the London Pub was attacked on June 25, eleven years later, I was glad I wasn't there. But the fact that we were deprived the opportunity to gather afterward meant that we had to stand alone in grief and fear, yet again."

Now Endestad works for the internet analysis company Analyse & Tall, which, among other things, investigates how arenas for hate speech form on the internet, what's required for good digital conversations to occur, and how democracy can work in social media. Her job is dedicated solely to creating good and inclusive communities.

Adrian Pracon also dreamed of being part of a strong community. After working in Skien for two and a half years, he went back to school, and his relationship with his family improved. He became involved with FRI and in student politics, and from there he was recruited by the Workers' Youth League. Pracon wanted the world to be a place where being openly gay is considered totally natural, a place characterized by diversity and inclusion. And these values were something the Labor Party could offer him. After that, everything moved very quickly:

he became the county secretary of the local Workers' Youth League, and that summer, he and seventeen other young people from Telemark would attend the organization's annual youth camp. They would be going to Utøya.

This weeklong political workshop is filled with an intense feeling of community and togetherness. And on July 22, 2011, the island was buzzing with a mix of young and old, city kids and refugees, established government members and promising youth politicians, all united around common values. It was a summer day of camping on an island in the fjord, and none of the people staying there felt in any danger—until just after five o'clock, when Anders Behring Breivik came ashore. This very community was what the terrorist was out to destroy when he began shooting.

After forty minutes of stalking and killing panic-stricken children, the terrorist came across twenty-one-year-old Adrian Pracon. But instead of any shot being fired, the two of them faced each other momentarily, the young man and the murderer. What did Breivik see in the crosshairs when he raised and aimed his Ruger? Because in that one decisive moment, Pracon could think of nothing other than his own funeral, he could see it clearly: his crying parents, his coffin. But then the terrorist lowered the rifle and continued on his way.

This moment of silence during the Utøya massacre is so clear and unambiguous, and so bizarre, because it represents a clear break from the twisted logic that had been applied until then: The blond young man, standing at the water's edge. The blond man, exactly ten years older, taking aim before letting the weapon drop for no discernible reason. The terrorist had sought to kill as many people as possible on that island, no matter who they were. So what was he thinking when he let this one young man go? What coincidence had prevented Pracon from being shot?

"Some people have a look that makes them appear more left-wing than others," Breivik said at his trial in 2012, where he was accused of murdering seventy-seven people, at Utøya and in Oslo. "But Pracon actually looks quite right-wing. When I saw him, I actually saw myself. I know I look like a right-wing person. I think that's why I spared him."

After waiting more than six months to get an answer to his question, and having countless sleepless nights in the weeks before the trial, Pracon had finally been given an answer. In all this targeted and deranged killing, the terrorist had momentarily *thought he had seen his own face* and been unable to fire. As though he had seen himself in a mirror.

But mirroring is only the start of having empathy; it's not enough to think that someone looks like you. Reflections always say more about the person being reflected. Being understood is about something more fundamental; it goes deeper. When we understand, we are part of the same flock. When we understand, we are no longer a danger to each other, we are part of each other's world. If we understand how others feel when they are afraid of something we are not afraid of, or love something we do not love, that something is no longer a threat to us but a natural part of the immensely strange and varied human herd. Understanding creates room for differences.

When Ninny the invisible child becomes visible again in Tove Jansson's book, it's not because of mirroring or copying the Moomins; it's a result of finding herself and her place and being warmly received. She experiences genuine understanding, the sort that enables a small herd to accommodate a variety of individuals with different needs. Accepting our differences is easiest when everyone in the flock is able to contribute to the community in their own ways. The human herd can handle many differences; it's what made us the most unique species on Earth: we are different and contribute in our own ways. We like cooperating and helping each other.

But it's hard to help those who don't ask for help. And it's hard to help those who don't even know what help they need.

When Adrian Pracon came home from Utøya, he felt like an outsider in an entirely different way. Few people understand you when you are battling mental illness, and it is also difficult to ask for help. When you are mentally ill (and in many cases this applies to physical illness as well), you will often feel ashamed of yourself, and that makes it hard to be understood. It is hard connecting with other people if you're living in constant fear. Like many other Utøya survivors, Pracon was suffering from PTSD, four letters that were totally unknown to him before Utøya, but whose meaning will stay with him throughout his life.

I have been to Utøya with Adrian Pracon. We found the place he stayed in for most of the attack, the same place where he was eventually shot—because after the initial face-to-face encounter, when his life was spared, he was nevertheless hit by one of the terrorist's bullets. Pracon had lain down under a raincoat and was playing dead among the lifeless bodies of murdered teenagers, when the terrorist returned and stood over him. Breivik was presumably aiming for Pracon's head, as he had done with most of those he killed, but it's possible that he couldn't properly see underneath the raincoat. He instead shot Pracon in the shoulder at close range, with a bullet that exploded into fragments and spread throughout his upper body. It was the last shot fired on Utøya before police surrounded the terrorist, who then gave himself up. By then—after being able to continue his rampage for seventy-five minutes undisturbed—he had shot and killed sixty-nine people, most of them children and young adults. In the center of Oslo, as the smoke cloud above the Government Quarter subsided, ambulance personnel retrieved the bodies of another eight people. Thousands more would experience the violence directly—as relatives, witnesses, and survivors—plunging them into trauma and grief that might never heal. A quarter of Norway's population knows someone affected by the July 22 terror attack, and since the bombing and killings, the entire population has become more afraid. There was huge concern about the possibility of more attacks, particularly among those living in Oslo and under the age of twenty-five. Even in neighboring Denmark, there was a noticeable increase in fear and psychological distress after the attack.

On Utøya, Pracon witnessed teenagers being killed, and as he hid under the raincoat near the island's southern tip, he also had to accept his own death. After experiencing something this traumatic, the external wounds, despite being the most visible, are not those that trouble you the most. The psychological damage after a sudden, traumatic event such as the 2011 Norway attacks is far, far worse. Research from the Norwegian Center for Violence and Traumatic Stress Studies shows that many people who weren't even working at the government offices on that terrible day suffered from PTSD. The reason they were traumatized, despite being at home or at the cabin when the bomb went off, may be that they knew some of the dead or wounded, and that they could clearly see how the bomb could have killed them had they been in the office that day.

PTSD can be described as a result of extreme stress. When you are very stressed, you function a lot in the autonomic nervous system, called *sympathicus*, which is supposed to ensure your survival in life-threatening situations. According to the American psychologist and neuroscientist Stephen Porges, the three stages of survival response in sympathicus are, first, an alert and quite communicative response, where you assess the situation based on your environment, as a child does when glancing at their parents to see if a toy is safe or not. (This has previously been referred to as "freeze" but is actually a form of vigilance.) If there is real danger, you then go into "fight or flight" mode—where most people will choose flight and only a few will fight. Finally, if you don't escape but still understand that you are in mortal danger, you go into a kind of full surrender called *vegetative parasympathicus*. Extreme fear can manifest itself during the so-called fright level, which can be described as a kind of emotional collapse where you lose control of your bladder and stool, vomit, and pass out. But this total submission when faced by an extreme threat has a clear function. If you're on the ground playing dead, any animal that isn't a scavenger will leave you alone.

All these ways of reacting to danger are entirely normal. But they are not functional when the danger has passed. When you have PTSD, your body and brain remain locked in a feeling of mortal danger. Sympathicus and vegetative parasympathicus are the nervous systems of loneliness, where you switch off communication with other people. And this works quite well, because extreme stress should only last for a moment. In pure survival mode, both your immune and digestive systems are paused while you escape to safety; you think short-term and not all that creatively; you take shorter breaths and get tunnel vision. It is a short-term solution that only works there and then. But if you remain in that state for years, it will take its toll on your body. Parasympathicus is the nervous system of attachment. But those with PTSD function in sympathicus, the nervous system for loneliness, which makes it hard to relax and thus difficult for them to connect with other people.

One of the terrible things about trauma memories is how easily they can be triggered. They can occur again and again, suddenly and without warning. And they do so in order to prepare us for future danger—we must never be surprised again. A trauma reaction can be sparked by the tiniest of things. New Year's fireworks can remind you of gunfire. The smell of burnt wood can evoke the aftermath of a bombing. A broken glass and a ketchup stain might remind you of bullet holes in a café window and blood running from a young body. You might be at a children's birthday party when these triggers occur, with tears rolling down your face—and your strange reaction to these seemingly harmless things will put you totally outside the fun going on around you. It's hard to understand the reactions of someone with PTSD.

This is precisely what Adrian Pracon experienced, once again finding himself outside the pack; suddenly, the killer would be standing there, at the local shop, for example, with a rifle in his hands—impossible, of course, because he was behind bars. But that's the nature of PTSD. It's like living in a horror film you cannot escape from, while everyone around you is living their normal, everyday lives—like two parallel realities, where the person living with PTSD lives in a frightening reality, separate from everyone else. It is dreadfully lonely. For a period after the Oslo attacks, every Norwegian newspaper had the gunman's face on the front page, which made it impossible to avoid seeing him, talking about him, being aware of him. And for the survivors and bereaved relatives, this became an additional burden. The images already haunted their nightmares, but he was also everywhere they turned, even when they were awake.

Pracon is now hypervigilant, always on guard, restless, and distracted. He cannot enter a room without looking for escape routes. PTSD can make you irritable and angry, it can cause anxiety and depression—and this makes it hard for the people closest to you to understand. Just talking about a memory can trigger vivid images. A sudden, unexpected smell or sound can make Anders Behring Breivik seem to materialize with a rifle in his hands.

The worst thing about PTSD is that the harder you try to escape the memories, the worse and more persistent they become. Trying to avoid everything that reminds you of what happened, to escape the ghastly horror movie you're constantly drawn into, is stressful in itself and becomes an additional trigger. It is a vicious and self-reinforcing spiral in which you isolate yourself with your fear.

Unable to sleep, Pracon was given powerful sleeping pills. But they didn't help either. Pills don't work when the body is in nonstop crisis mode and pumped full of adrenaline. He was also offered psychological help, though he wasn't sent to a psychologist who understood PTSD but to a local couples therapist. He began losing his grip on reality and feeling genuinely suicidal, while people all over the country sat with psychologists and priests and paramedics and told them gut-wrenching stories about friends who were killed and children who were gone forever. There were policemen who described how the dark night on Utøya was illuminated by the flashing screens of abandoned telephones, lost by their young owners who were now dead but whose parents refused to stop calling. There were nurses and government workers and security guards and random passersby and people who mopped up the blood in the café, and so many more. And very few of these people got the help they needed, if they got any help at all. It is lonely being assaulted by terrifying memories, and it is lonely being in perpetual crisis mode in one of the world's safest countries, because no one around you can understand what you're talking about; even the country's psychologists—most of them, at least—were at a loss.

There were 546 people on Utøya, and if each of those had ten relatives, which is a reasonable estimate, it means that six thousand people were directly affected by what happened there. In addition, there were two hundred people injured in the Government Quarter, plus the 250 people who were in the government buildings when the bomb went off, and the seventy-five people who were in the immediate vicinity and could also have been killed, plus the five hundred people who were in the center of Oslo, which makes 10,250 relatives of the attack in central Oslo. We also have all the people who were near the island at the time, all those who drove their little boats out to save the escaping teenagers while being shot at by Breivik, all the police officers and aid workers, and once again all their relatives… Altogether, it's no exaggeration to talk about over twenty thousand people (and there could be many more, of course), relatives and immediate victims who will be deeply affected by the actions of a solitary, lonely perpetrator, every single day, for the rest of their lives. Many of them have developed PTSD—and many live with so-called complicated grief, a strong and debilitating grief that lasts for many years.

"That unnatural death should be considered a public health problem is supported by large international registry studies which show that mothers who lose children will have an increased risk of early death for the rest of their lives," write psychologists Kari Dyregrov and Pål Kristensen, who have studied the bereaved families of Utøya victims. "Fathers will have such a risk in the first years after the child's death. Serious health consequences such as an increased risk of cancer and heart disease, high alcohol consumption and suicidal thoughts, have also been documented in those struggling with complicated grief."

Dyregrov and Kristensen found that of the parents whose children died on Utøya, 82 percent have complicated grief, and 63 percent suffer from PTSD. There are also those who live with grief and trauma for other reasons: survivors of house fires and road accidents, those who have experienced rape, abuse, and childhood trauma, Norwegian war veterans who served in Lebanon and Afghanistan, anyone who has experienced a sudden death in the family. All the bereaved following a suicide, which is now one of the most common causes of death among teenagers and young adults. There are hundreds of thousands of survivors who live with grief and trauma every day, and those who have experienced something so tumultuous will often feel very alone in that experience. Research shows that it is easy for these people to believe that no one can understand them, and as a result they withdraw from the community, they isolate themselves. Grief and trauma create a paralyzing loneliness, because it is so difficult for others to understand how you feel.

Eventually, Adrian Pracon received proper help. He initially asked to be admitted to a nearby psychiatric hospital. But his application never left the GP's office, and in the end, he became so desperate that he simply turned up at the local psychiatric-emergency unit and told them he needed to be admitted, otherwise he would have to kill himself. And there he was finally given treatment that helped, a therapy based on eye movement called EMDR. No one knows why EMDR works, but it involves following a moving light while processing the trauma memories. Somehow, these eye movements recode your memories and make them less charged with extreme fear.

"When I found a therapist who had lots of experience in trauma and EMDR treatment, I finally got the help I needed. She was able to help me," Pracon says.

But this only happened after an incident where Pracon got drunk and punched two people to the ground outside a pub in Oslo—an episode he has no recollection of, but which nevertheless landed him in court. This outburst of violence came after a long period of PTSD-related stress, but he is now very careful with alcohol, since he never quite knows how it will affect him.

"I have to write the instruction manual for how I work as I go, all the time, always," he says. "There's no blueprint for it."

Staying as mentally healthy as possible is work, and lonely work at that, because no one else can do it for him, and no one else can fully understand what the work entails. Living with trauma memories means that days can be both good and bad, but hard to prepare for. It's impossible for him to know just how connected he may be to his surroundings, how little space his inner world will take up that day. When his sleep worsens and is more intermittent, it weakens his resistance to negative thinking and sends him down, in a vicious spiral, into a maelstrom of darkness.

When times are bad, your inner life can take over completely, and the external world, people, and impulses become less clear. Your brain goes into survival mode, while your body's stress response goes into high gear. When we are very stressed, connecting with other people is almost impossible.

"It's important to get myself out of a negative episode in time," Pracon says. "Otherwise, I end up physically exhausted and my mind wanders for ages. I've been on sick leave for many years; I know how far down you can get. When it was at its worst, I felt suicidal."

Spiraling thoughts can lead to sleep problems, stomach pain, muscle pain, suicidal ideation. Some days, neither his body nor his head will work, which means that when making plans there are two things Pracon must consider: everything he wants or needs to do in life, and the trauma. He could, for example, optimistically plan on taking an exam, but he would be doomed to fail should the preceding weeks be plagued by sleepless nights, trauma memories, pain, and fear. Pracon did, however, gain a BA in peace and conflict studies and a master's degree in international security from Oxford Brookes University. While living in Oxford, he also met Jack, whom he now lives with just outside Oslo, in a large house filled with plants and a three-year-old husky who was rescued from the streets. Today, life is pretty good for him.

Pracon had planned on using his knowledge of terror in the fight against terror, but his hard-earned experiences in acquiring good mental health, health care, and political work led him in another direction. The most painful thing in his life has become his strength, something he can use to help others, and he is now a political adviser in the organization Mental Helse [Mental Health], where he lobbies to put mental illness on the political agenda. Throughout the pandemic, he saw how trauma, depression, and suicidal thoughts contributed to bursting the capacity of the Mental Helse switchboard; he has seen that loneliness in Norway has increased. At the same time, he has to work with his own PTSD on a daily basis.

"It wasn't a conscious decision, but I isolated myself from the Workers' Youth League afterward. I wanted to put all the pain behind me, of course, but it meant that I withdrew from the only people who really understood how I felt and what I'd been through," he says.

The people most able to understand him were the ones triggering memories of the attack. Each and every face was a possible door into the darkest experience of his life.

So in order to become part of the herd, Pracon had to come up with a strategy that would require a lot of courage: he would change the herd. Changing himself was inconceivable: he couldn't stop being gay, and he couldn't stop having PTSD, which meant the world had to become a place that accommodated him. This herd, which had the potential to exclude him, would have to become a slightly different one.

He came out as gay and got involved in FRI, the Norwegian Organization for Sexual and Gender Diversity. He wanted the world to be a fairer place, so he joined the Workers' Youth League. He is a victim of terrorism and wants to prevent terrorism. He suffers from PTSD and wants more openness about mental health. It seems to me as though he is trying to make the world and himself play the same tune.

"Yes, that's what I want!" he says.

I think of the word *osmosis*, the chemical process where two liquids attempt to balance each other. For example, two quantities of water, one saline and the other fresh, are separated by a membrane that allows the two liquids to seep through. Over time, they compensate each other and try to achieve an equilibrium. They will not be kept separate; they want to be mixed. Maybe people are like that. Maybe Adrian Pracon is like that. The man who had stood opposite him with a rifle had so thoroughly misunderstood him: Adrian simply wanted to be understood for being different, he wanted to make the world a better place for himself, he was fighting for a world where there was room to understand people's differences.

"I feel like I have a responsibility to talk to the terrorist. Perhaps something good might come out of it? He might give me more of an explanation. After all, he targeted and spoke to me directly before. I want to know what happened to him as a child, what kind of trauma he is carrying," says Adrian. "I want to understand.

"I have been angry. But at some point I decided that I had to put the hatred aside—that I had to accept the situation, because the alternative was making me more ill," he says.

"I feel like I'm afraid of triggering you when we talk about trauma," I say worriedly.

"You don't have to be afraid of triggering me. I get triggered no matter what, by all kinds of things," he replies reassuringly. "It's something I rarely talk about specifically. Because I don't want to be different. I'm so tired of being different! I couldn't be a normal straight man. Now I can't be a normal gay man either, because I have PTSD, and I'll have to carry that for the rest of my life. I'll always be affected by what happened. I avoid talking to my friends or my boyfriend about it, because I don't want to be 'that guy,' that sick guy. It can make me feel very lonely at times."

I know a bit about how Adrian feels: many years ago I, too, was diagnosed with complex PTSD. This means that PTSD has characterized my whole personality, from an early age. It's made me irritable and jumpy, given me sleeping problems and eating disorders, and I've smoked obsessively to ease the turmoil within me. The profound restlessness driving me makes it hard to connect with other people; I'm always trying to get as far as possible from any potential source of pain.

I lie about how I feel. Most people see me as outgoing, relaxed, warmhearted—a person with plenty of energy. Behind the facade, I feel like I'm barely surviving, like I'm scraping my way through life by my fingertips. If there's a conflict at home, I often find it almost impossible to deal with, everything falls apart, the anxiety grows and becomes so overwhelming that I have to leave the apartment, my sleep deteriorates, and the knot in my stomach gets bigger. On a good day, life feels wonderful. But there are very many bad days, when I become like Toffle barricading himself indoors and like the Groke howling in the night. And during the pandemic, bad days were the ones I had most.

When my father died suddenly in 2022, this turmoil grew into a full storm. He died of Covid, just as the pandemic was being declared over, when the world was opening. On the quiet side street we live on, late night revelers were once again breaking the silence as people celebrated the end of the pandemic, but I was in a totally different place mentally. There seemed to be a party every single night that spring, but in our house, things were very gloomy. I couldn't look at my own feet, without bursting into tears, because they reminded me of my dad's; I remember us both putting them side by side when I was a little girl, mine small and smooth, his big and hairy, and me saying, "Look, we've got the same feet!" Now I can suddenly cry at the sight of my feet, and it will be impossible for most people to know why.

A poem by Kjersti Bjørkmo that was published just before the pandemic, in 2019, has stuck with me for years. It is about community. But I've now realized that more than anything else, it is also about grief—about grief that is especially hard to bear when everyone around you is preparing for a national celebration, in this case Norway's Constitution Day:

A warm night before the seventeenth of May after a death
    in the family.
The window is open wide, facing the road leading into the
    forest.
I stand for a while to see if the resident badger might saun-
    ter past.

I so want to see it walking quietly between the lampposts
    with its long, strange face.
And soon it will be light. Soon I will hear the sound
    of dress shoes upon asphalt.
Soon push chairs and plastic whistles.

Irreplaceable, those who awaken. Those now stretching.
Those going barefoot into the kitchen and making coffee.
Irreplaceable, those sending text messages. Those running
    for the bus, spilling ice cream on their shirts.
Those losing their keys on the stairs.
Talking quietly in a hallway.
Irreplaceable, two siblings
    who take turns carrying a sponge cake through the city.

It's a poem I repeatedly come back to. Because at first the
narrator seems totally disconnected from the outside world.
Someone has died, she is grieving. She is waiting for the lonely
badger, but she is instead given everyday life, the chaotic spec-
tacle of the May 17 children's parade. The poem's ending makes
me cry almost every time. All the people, filling the streets with
plastic whistles, running for buses, have become so mortal suddenly. Death is everywhere. The small things we do in life have
become so vulnerable. Maybe because that's how it feels, living with other people, the whole human project: I picture the
cake—an almost obligatory item, transported between families' and friends' houses before the parade—sinking on its way
through town, tilting sideways, its cream souring in the heat of
the sun. It won't be perfect, but we try: we uphold our culture
and understanding, we uphold our words and dreams between
us, while carrying an entire sponge cake through the city. And

in this great project that is humanity, we are irreplaceable, all of us, uniquely strange. It won't be perfect, and we can't understand each other completely. But most of us do try.

5

Why community can be dangerous

SEVERAL DAYS AFTER the Utøya attack, something incredible happened. A total of one million people—a fifth of Norway's population—gathered across the country, with roses in their hands, to remember the victims. On July 25, 2011, just three days after the attack, two hundred thousand people took part in one such "rose parade" in the center of Oslo; it was the biggest commemoration in the country's history. I was there, shoulder to shoulder with everyone else. And we were all aware of one thing, it was impossible not to be: as we stood outside Oslo's town hall that afternoon, we all knew that if Breivik had sympathizers or even accomplices, this was a golden opportunity to follow up on their previous attack. It was as though we had made some kind of island, our bodies huddled together, a circle of light and belonging, beyond which all manner of terrifying things could lurk. I was afraid, just as many others probably were. But for each person who stood there holding a rose, there was something else, something even more important: we wanted to stand together against the hate. But what hatred and racism really are wasn't discussed very much.

A test developed by Daniel Russell in 1978, the so-called UCLA Loneliness Scale, measures how connected we are to our communities. It consists of twenty questions and mainly concerns the innermost circles of love and friendship. It also measures *social* loneliness, which is a sense of not having a place in society, of not belonging. "Social loneliness occurs when a person lacks social integration in society or is not involved in society through friendships, neighbors and colleagues," wrote loneliness researchers Daniel Perlman and Letitia Anne Peplau in 1998.

People living in societies with very strict hierarchies are more prone to depression, as is shown by the famous Whitehall Study, which, from 1967 to 1977, examined the mental health of London's eighteen thousand civil servants. What the study made clear was that a lack of access to social mobility, control, or creativity increases depression among those at the bottom of the hierarchy—where you aren't noticed, where you are constantly overlooked, where you know you are insignificant. We humans aspire to climb within the hierarchies we relate to and become depressed if we fail to do so, which may again be due to our fear of rejection. If you are near the top of a hierarchy, you're presumably more likely to be protected and cared for. And if you feel unimportant, have no leeway, and are defined as one of society's losers, you withdraw in shame. If you don't feel like you are supported and cared for by the community, and don't share a common goal with your community, you become lonely. You give up.

"Social isolation deprives us of both our feeling of tribal belonging and our sense of purpose. On both counts the effects are devastating, not only for the individual, but for societies as well," writes the researcher who has been most important for the understanding of loneliness, professor of neuroscience John Cacioppo.

Loneliness both deprives us of the herd and destroys the herd. And this feeling of outsiderness doesn't occur without reason; it is a direct result of small and large exclusions that are aimed not only at individuals, but at entire groups. It is clear that some people are far more likely than others to fear ostracism—and thus loneliness. I think we just lack a word for it: *to lonelify, to inflict loneliness on someone.* Some people are lonely more often than others. It's not difficult to understand why there can be more loneliness in a society that puts success and achievement before everything else, that celebrates winners, where the preferred body is young, white, strong, and skinny: just look in the corners and you'll find them, those who *haven't succeeded*, who fear ostracism, who have been pushed to the fringes. You'll find the lonely among the poor, sick, and old; among people with disabilities; people with a body, skin color, sexual orientation, or gender identity that doesn't fit the "standards" the community has decided are correct; among immigrants and refugees, the mentally ill, people who are grieving or struggling mentally, those who take drugs, those in prison, school dropouts, and daydreamers. It is these people who fall outside the pack as a group and who also experience greater loneliness as a group.

When entire groups are invisibilized, it erodes society from within, and those concerned get eaten up from within. And while it sounds very dramatic when I put it that way, there is research on this very subject.

In the late 1970s, Princeton student Arline Geronimus discovered something unusual: she observed how poor African American teenagers suffered from a variety of diseases, such as autoimmune diseases, diabetes, and overall aging. Melanin-rich girls aged physically more quickly than their white friends. Geronimus later became a professor at the University of Michigan Population Studies Center, where she continued to work on unraveling the mystery of why this happened. As she gathered data, it became clear that what she had seen wasn't a coincidence. She called it "weathering," referring to how stone is gradually worn down over time. Although people were skeptical of her weathering theory when she first launched it, subsequent research into chronic stress and what it does to the body has proven her right. Environmental factors such as poverty and racism do break down the body, causing low intensity inflammation, which in turn causes heart and vascular disorders, diabetes, rheumatism, autoimmune diseases, cancer, obesity, and a range of other symptoms that we recognize from loneliness research.

"But what I've seen over the years of my research and lifetime is that the stressors that impact people of color are chronic and repeated through their whole life course, and in fact may even be at their height in the young adult through middle adult ages rather than in early life. And that increases a general health vulnerability—which is what weathering is," Geronimus says today, having built up a solid research environment around her. "One reason people dismissed it is that I first observed that young black women were more likely to have poor pregnancy outcomes if they were in their mid-twenties than if they were in their late teens. And this flew in the face of a lot of advocacy organizations that were working very hard to prevent teen childbearing."

For women in general, having children in your mid-twenties is definitely an advantage, and teenage pregnancy has major health risks attached. But the research shows that for melanin-rich women in the 1970s, having children in their teens was better, because it meant that the physical aging had not yet progressed.

Of course, it was often argued that this was because of genetics, so Professor Geronimus had to first examine her research subjects' genes. It was claimed that African genetic material was somehow more vulnerable to modern American life. This turned out to be wrong. Nor was any of it linked to lower education, because even middle-class African Americans are subjected to this weathering. Disease and premature death were being caused by something else entirely.

Research on inflammation has accelerated in the last few years, as has knowledge about epigenetics. The results of long-term, low-intensity stress are deposited very clearly in the human body, right down at the telomere level. Telomeres are protective regions at the tips of our chromosomes—the structures of DNA within all our cells—and they have an effect on our longevity. They are biomarkers of biological aging. After looking at all the possible factors that contributed to aging in the bodies of African Americans—yes, why these people had shorter telomeres—Geronimus noticed one thing that stood out: their bodies were being broken down from within by *racism*.

A large study published in 2018 examined Geronimus's hypothesis more thoroughly: researchers at the University of Georgia followed four hundred African Americans over a twenty-year period to see how discrimination affected them in terms of elevated stress levels and low-intensity inflammation. They took seven measurements during this period, and what became clear to them was that discrimination is most harmful during childhood.

"The bad news from our study is that early exposure to racism produces lasting effects on an individual's risk for elevated inflammation," writes the research group.

"Our findings show that weathering starts early in life and continues to be harmful to health into adulthood," they state.

So what exactly is racism? What does it mean to be told through looks and laughter, body language and words, jokes and stories, through invisibilizing, humiliation, and violence, through inadequate police protection and worse treatment from the health care system, that you don't belong, time after time, year after year? If loneliness is *the fear of being ostracized from the pack, with a potentially fatal outcome*—wouldn't being subjected to racism trigger a powerful loneliness response? After all, you're being told that you don't belong to a herd, simply because you have a particular skin tone! What is racism if it isn't numerous minor and major threats of ostracism, and what is experiencing racism if it isn't pure loneliness? Perhaps loneliness isn't a mysterious phenomenon that suddenly occurs, but a totally logical response to a series of minor and major exclusions that follow a specific pattern.

If you experience a lot of discrimination as a child, you become more sensitive to it and become more shaped by it. A child's brain is very plastic, and childhood is when you learn strategies for how you relate to other people. You are especially sensitive to whether you belong or not, because your herd is even more crucial when you are vulnerable and weak and totally at the mercy of your surroundings. A child comes into the world naked and knows nothing, and soon, within a few years, some children know that they are less valuable than others, a less accepted part of the herd than others. And it changes them: it changes their bodies and their expectations of the world.

"Of course racism makes you feel lonely! What the racist is conveying to you is that you are an inadequate human being, so theoretically you can be killed," says Guro Sibeko, who has written *Rasismens poetikk* [The poetics of racism] and *Alle blir fri* [Everyone will be free].

"Someone might make a random comment at a party, while smiling and surrounded by people, and if those people don't realize this is a threat, it will reinforce your sense of loneliness and alienation, because they didn't respond," she says. "Many victims of racism see themselves as weak: as a child, you have to just think that adults who talk like that are quite stupid, but that's scary, to think that you're being cared for by people who are stupid. So you end up accepting their worldview instead. Nearly all children who experience racism choose to distrust themselves. They become used to not trusting their own instincts, which in turn means they are also less capable of looking after themselves."

Racism belongs to the outermost circle of loneliness—aimed at a group but felt on the body by everyone subjected to it. You can live within the four walls of your home, with close bonds and good friends around you, and still experience this kind of loneliness. It is one form of loneliness: the loneliness of not being considered a full-fledged part of the society you live in.

Feeling excluded from the grand narrative leaves you with a sense of intense outsiderness. And overcoming this particular form of loneliness is difficult. As the Norwegian journalist Yohan Shanmugaratnam has experienced.

"I have a role, and a relative amount of power," he says. "I can write and say what I want, I'm a man and not a Muslim, so I'm privileged. At the same time, I know that I don't belong one hundred percent in this society, simply because I came here when I was six years old, and I still have a nostalgic relationship with Japan. I've always known that I came from this other place." Shanmugaratnam was born in Sri Lanka to a Japanese mother and Sri Lankan father. His early years were spent in Japan, but he grew up in the small and safe community of Ås, outside Oslo.

In his book *Vi puster fortsatt* [We're still breathing], which is about racism and the July 22 attack, he describes what it's like if he catches the bus and sees an empty seat beside a light-skinned woman. He writes that, rather than risk scaring the passenger, he'll choose to remain standing. He adopts a lonely person's behavior—he withdraws.

"I've heard what it's like from other people, and even the most bullish gangster teenagers can withdraw similarly," he says. "And it's connected to often being asked 'where are you from?' even though many of us were born here. You are constantly reminded that you are different. There is something universal in the experience of being born here while also belonging somewhere else. You don't belong one hundred percent to this one place, even if you do officially. So even those who come from here can feel like they're not from here."

The sense of being outside and not quite part of the community is insidious and indistinct. It's hard to be sure what's what. I wonder how it feels when people talk past you, about you, on behalf of you, like you're not there. How does it feel if you're never actually included?

The author Amiri Baraka (who became a Muslim and changed his name from LeRoi Jones) was educated at Howard University, where he only read literature written by white men. In the 1960s he became an anti-racist and activist writer. "I was taking words, cramming my face with them," he wrote. "White people's words. Profound, beautiful, some even correct and important. But there is a tangle of non-self in that for all that. A non-self creation where you become other than you as you."

In a poem from 1964, he describes how he sees himself with the white man's gaze, and how this gaze has become fully incorporated into the way he thinks: "I am inside someone / who hates me. I look / out through his eyes." It shows how complicated knowing yourself can be.

"What's hard to know is how much of my self-perception is drawn from the way immigrant men are described in societal debate, and how much is linked to my own personality traits, because I'm quite an introvert," Shanmugaratnam says.

He has always known that he comes from somewhere else, while simultaneously trying to consider himself Norwegian.

"I've experienced a lot of inclusion," he says. "But when older people asked me where I came from, and I didn't have a clear answer, since I come from three countries, they would just end the conversation by saying 'yes, but you're Norwegian.' They probably meant well, but at the same time it made me feel like they couldn't deal with the ambiguity. I interpreted it as meaning that there can only be two answers: You're either Norwegian or you're not."

It's a feeling of sadness, of not quite knowing where you belong, a rootlessness that is part of you, and an uneasiness that becomes permanent.

"I never belonged to the herd in the first place," he says. "It's the eternal melancholy of the migrant. It manifests itself differently in each of us: sometimes as restlessness, sometimes as an identity crisis and fanaticism, while for the vast majority it is just something vague and undefined that we carry our entire lives. It doesn't have to be a problem either, but it's a condition we may never be able to put words to. I'm trying to understand myself, and I kind of don't quite know where I end, and where all the cultural impulses I've received begin. What am I?"

Vi puster fortsatt [We're still breathing] begins with a beeping noise, "like the tone that used to come from the TV when broadcasting ended for the day and the test card appeared," he writes.

Shanmugaratnam had just walked through the Government Quarter with his two children in a stroller and was about two hundred meters away when the explosion went off. This meant he had walked right past the white transit van containing the one-ton bomb that had been parked there at 3:16 p.m. that day and had been there for nine minutes before detonating. It also meant that for a few minutes, he was in roughly the same place as my friend Ada. They would have passed each other, or at least been only meters apart. Ada called me shortly after the blast. She was frightened and upset. And while the shock spread through the city, through the country, Yohan stood in the nearby square and saw the black smoke billowing from the government buildings and the office documents fluttering to the ground.

Yohan Shanmugaratnam's book is about racism and about Breivik's attack on the defenders of multicultural values. It was no coincidence that the ruling Labor Party and its youth organization were attacked that day. They were so-called cultural Marxists, and considered traitors by many on the far right. Their crime was wanting an open and inclusive society, a society where no one should feel left out and lonely.

The struggle against racism became more visible after the murder of George Floyd in 2020. Black Lives Matter expanded and became a huge global movement. Millions of people demonstrated around the world.

"Black Lives Matter broke down a lot of invisible barriers," Shanmugaratnam writes, "and it suddenly felt like we were in an open-plan office and could stand up for each other: Now I know that there are other children, who, at other times and in other cities, have tried scrubbing their skin color away too. It means your experience is no longer a personal one, but part of a bigger picture. Because trying to scrub your own skin color away—that's the epitome of loneliness!"

There is a multibillion-dollar industry that sells skin bleaching treatments to melanin-rich people. Millions of people around the world bleach their skin to become whiter, according to the World Health Organization, which estimated the industry would be worth $31.2 billion by 2024. These products can cause liver damage, kidney disease, cancer, neurological problems, and stillbirth. Yet people are willing to take the risk, to erase their own heritage and history, to become an unnoticeable part of the white majority. Shanmugaratnam believes that what happened with Black Lives Matter highlighted a widespread problem: how racism acts as a kind of poison.

"It's a feeling of not being valued as a human being, and while you don't quite understand it as a child, you still know that it's wrong. So: Are we being overly offended and woke, or are we just tired of our human dignity being systematically underestimated?" he asks.

Shazia Majid, a columnist at the Norwegian daily newspaper *VG*, always knew that some people weren't counted in the big "we," so it was those people she wanted to write about. She wanted to write a book about the people who are never mentioned in history books, who are written out of the story of Norway.

"I wanted to tell my mother's story, the story of her life, because it was that life I was familiar with, and it was so dramatic—a truly heroic story!" she says.

Majid—whose mother came from Pakistan to join a husband who had already moved to Norway—grew up in a strictly patriarchal structure. Her mother wore herself out through hard

physical work and lived almost entirely outside Norwegian society and without any network, though she was one of the lucky
ones: she had a job and a Norwegian friend. The women who
came from Pakistan as migrant workers in the 1970s were never
included in the nation's story, and perhaps not even their own:
those who worked had terribly paid jobs, few of them learned
the language of their new homeland, and they were subjected
to racism and exclusion. They didn't know their own rights
as mothers and workers; they didn't understand the culture,
the food, and the habits. Imagine moving to a country where
you don't understand the language or the customs, where you
live in total isolation, are looked down upon and don't make
any new friends, where even your own children have a better
understanding of the culture than you, where you have to work
several jobs and can still only afford to live in a cramped and
dilapidated home. All this while being considered some kind of
possession by your own father or husband.

"Pakistani women then were living the kind of life that
Norwegian women had lived a hundred years earlier. While
Norwegian women were fighting a battle in the public sphere,
Pakistani women were fighting the patriarchy at home. It was a
colossal difference. Immigrant women weren't seen; they were
considered an appendage to their husband, by everyone, both
inside and outside the home," says Majid.

Majid is a highly educated divorced woman, with a prominent job as a national newspaper commentator, who is also
Muslim and the daughter of Pakistani parents. So her whole life
concerns reconciling opposites. Her commentating focuses on
understanding differences and reducing conflicts across genders and cultural divides. In her own way, she attempts to build
bridges and patch society together.

"My aim has been to foster understanding and communication," she says. "And I've now learned to speak in a nonconfrontational way, in order to win respect and make my voice heard on both sides of these conflicts and battle lines. The worst thing about patriarchal structures is that women are not given the ability to express themselves or argue their case. But women aren't the only ones trapped by these structures; anyone involved in them becomes unfree."

In the strict hierarchy that a patriarchal system creates, both women and men are locked into roles that stifle their individuality and independent thinking; it is a straitjacket that prevents freedom of movement. And when women are not considered individuals, but men's property, they also become victims of objectification, rape, violence, and murder.

"The pandemic made it clearer to me who the most vulnerable are," Majid notes. "It has reinforced the inequalities, which are now just bubbling to the surface, so we have to keep an eye on the loneliest and most at risk. There are women and children exposed to violence, drug addicts and psychiatric patients, and low-paid immigrant families living side by side. And the women who were neglected when they first came here—my mother's generation—are still invisible. Their bodies are exhausted, and they have psychosomatic complaints and disorders. They are still, very many of them, extremely lonely."

She believes that the isolation they have experienced has worn them out, that the loneliness and invisibility has made them prematurely old and ailing. They have lived in a society that didn't understand them, that didn't see them. Many people experience the same thing: being considered less important. Racism and discrimination take many forms, but often manifest themselves similarly, as the same invisibilizing, contempt, and threats.

As I stood in the crowded square outside Oslo City Hall on July 25, 2011, with a wilting rose in my hand, it seemed totally imperative: we needed to find out how it could have happened—how one of us, a seemingly ordinary man from Oslo, could decide to kill politically engaged youths at a summer camp on an island. Were there more people like him? Who were these racists, and where was this hatred coming from? Who could be so hell-bent on removing whole groups of people from society that they were willing to commit mass murder? In the days following the attack, it was perhaps the only thing we discussed, and with increasing astonishment. In 2011, very few of us knew that behind closed doors and in corners of the internet, right-wing extremism was on the rise. The attack came as a surprise to many of us. It was like we'd been playing happily in a safe, green park and a huge crevasse had suddenly opened up in the middle of it. Many of us demonstrating in the "rose parade" no doubt believed that he was a lone wolf, an extreme one-off. Perhaps there were a few more, maybe ten? No more. There couldn't be more than that.

"Norway has passed the test. Evil can kill a man, but never conquer a people. Tonight, the Norwegian people are making history. With the most powerful of all the world's weapons—free speech and democracy—we are setting the course for Norway after 22 July 2011," said the country's then prime minister Jens Stoltenberg, with real conviction, that day from the stage outside Oslo's city hall.

Stoltenberg knew many of those killed or wounded personally. Yet he was totally calm onstage, despite being an easy target for a bullet as he towered over the mass of people.

There was this strange feeling that if we stood together just long enough, with roses in our hands, life would return to how it was before—that words and roses would be enough to reclaim the country we knew to be safe and good. The sociologist Émile Durkheim describes this type of gathering as almost electric. Rallying around clear symbols, such as roses and the word *democracy*, is intensely unifying; it becomes a core of the community that you can later apply to the trivialities of everyday life. "The very fact of assembling is an exceptionally powerful stimulant," he writes. "Once the individuals are assembled, their proximity generates a kind of electricity that quickly transports them to an extraordinary degree of exaltation."

Emotions were running very high that Monday—it was a mixture of grief and vulnerability—and this huge mass of people gradually but spontaneously began singing the national anthem, without anyone knowing who initiated it. This was followed by another standard, "Kringsatt av fiender" [Surrounded by enemies], based on the poem "Til ungdommen" [To the youth] by Nordahl Grieg, which I have never sung with so much emotion: "Those who with their right arms carry a burden / Precious and irreplaceable, cannot murder," we sang, Grieg's words carrying a thin vein of hope. "We will preserve the beauty, the warmth / As if we are carrying a child tenderly in our arms," we continued, two hundred thousand of us, crowded together in the giant square.

"Tonight, the streets are filled with love," Crown Prince Haakon said hopefully from the stage.

I remember how I cried when he said those words, and how I continued to cry, almost helplessly, because what use were speeches, really, when seventy-seven people were dead—and I hugged strangers and clutched the stem of my little yellow rose. The streets may have been filled with love, but I now knew that out in the night there were people with hearts full of hatred, planning their next attack. And there were more attacks.

Because here's the paradox: We may have promised one another, in the days that followed Breivik's attack, that we would fight right-wing extremism and racism, that we would oppose and expose all the internet trolls. But what actually happened was that right-wing populism and radicalization slowly but surely increased. Right-wing extremism went from being totally marginal to gaining more space. On August 10, 2019, Philip Manshaus killed his own sister, seventeen-year-old Johanne Zhangjia Ihle-Hansen, at home with four shots, before driving, heavily armed, to the Al-Noor mosque in Bærum, which he stormed. He had been inspired by Anders Behring Breivik and by Brenton Tarrant's mosque attack in New Zealand.

Journalist and author Lasse Josephsen knows a great deal about the right-wing extremism that develops in boys' bedrooms and was less surprised than most by the Oslo attacks. Josephsen has spent over twenty years investigating the darkest corners of the internet—places like Telegram, Gab, BitChute, Odysee, Nazi forums, 4chan, and 8chan, where people can anonymously post illegal material, such as videos of rape and murder. These online forums are where many people become radicalized. Josephsen has met some of them, young men who have fallen out of Norwegian society and a lot of Americans. The Americans are often men who lost everything during the 2008 financial crisis, or young boys from the lower classes, descended from generations of traumatized war veterans, who have lost all hope of a better life. They are poor; they have no future. Accelerationism—a form of right-wing extremism that idealizes an apocalypse that sweeps away the current social order—is a growing ideology, whose supporters actively seek to bring about catastrophe. They believe that after the chaos and carnage, a new world order will arise, a neo-Nazi one. It is the height of cynicism, and it prevails within far-right online forums. Fortunately, perhaps, right-wing extremists are so bad at building community that in many cases they don't pose any threat; they are so suspicious of each other and so disorganized that any attempt to organize a violent global revolution tends to implode. As we know from research, negative humor and bullying lead to loneliness, not to strong communities. But such like-minded groups, despite being illusory, are still communities.

"What people don't realize is that there are so many of them, so very many, and there will be more!" Josephsen says. "It's no longer on the fringes. And online right-wing extremism is born primarily out of loneliness and alienation."

Accelerationists themselves *are* still on the fringes, even within the far-right movement, but far-right ideas are beginning to seep into the mainstream. When he realized that the perpetrator of the Oslo attack was a white right-wing extremist, Josephsen's heart leaped into his throat: What if this was someone he had been in touch with? He was, after all, mixing with a large global underground, where people for years had openly flirted with far-right ideas and counter-jihadism.

Josephsen wasn't the only one who was scared. As the country realized that a major terrorist attack had taken place on home soil, many people no longer felt safe, including Norwegian Muslims.

"I was terrified . . . immediately after, when we didn't know," says Shazia Majid. "Those hours . . . I was eating dinner with a group of friends, all Muslims. And we couldn't eat. We were all afraid. We didn't dare look at each other. We tried making jokes to lighten the mood, but nobody laughed. It was a terrible feeling."

Shazia knew that Norway's entire Muslim community would become targets of harassment and perhaps violence if the perpetrator turned out to be Muslim. In the hours after the bombing, when everything was unclear and no one knew who was behind it, many assumed it was Islamic terrorism.

"That feeling disappeared the moment we heard the name 'Anders Behring Breivik,'" she says. "But that was 2011. Since then, being a Muslim has been a bit of a roller coaster. Especially around 2014–2015 with all the terrorist attacks in Europe, and with the Syrian refugees ... It was a time of constant anxiety."

There were quite a few Muslims who didn't feel welcome in the rose parade; many of them had just days earlier feared the worst. And they will continue to fear the worst if today's polarization and hatred increases. Polarization is something that infects every aspect of social debate; we become more focused on what separates us than on what unites us. Culture wars tear friendships apart and feed the online trolls. The standpoints become more intransigent. But when an apparently radicalized Islamist attacked the gay bar London Pub on June 25, 2022, Norway actually had its first fatal attack by a Muslim perpetrator.

"Norway has changed since 2011," wrote Shazia Majid in the newspaper *VG* immediately after the London Pub attack. "All the mass murders we've seen have been carried out by ethnically Norwegian men, until now. This has taught us several things, one being that terrorism isn't the sole preserve of Islamic extremists; that it can just as easily be carried out by a white racist, so it's best not to jump to conclusions. In that time, we've also had an open, sometimes painful, debate about immigration, integration, Islam and Muslims."

She believes that we've moved closer to understanding each other, not further apart. At least, that's what she hopes.

But that's not what Lasse Josephsen is seeing. The websites he sees, which were once filled with irony, memes, and transgressive humor, are now full of violent fantasies.

"Many of the young men in the forums are simultaneously funny and malicious," Josephsen says. "The gamers among them may be struggling at school, but they're kings in the gaming world, and on the forums some of these gamers feel like they can do whatever they want—and they're not particularly interested in political correctness."

The darkest corners of the internet are frequented mainly by young men who are falling out of the system, who feel that there's no place for them, that they don't belong and have no hope for the future. And for some, the exclusions and subsequent loneliness leads not only to depression and withdrawal from society, but to violence. The hedgehog puts up its spikes and hisses. The loneliness becomes aggression. An "attack is the best defense" reaction. As ostracism researchers have described, ostracism increases the risk of violence and mass shootings. Fear of being rejected by the community can make a person reject that same community through violent means. And on the right-wing-extremist websites, violent content and glorification of violence is the norm.

"Have you seen murder on these forums?" I ask Josephsen.

"Plenty of times, and sometimes it's absolutely horrendous to have experienced something like that, and sometimes I'm just totally numb," he replies.

Josephsen eventually became involved with the scene around Gavin McInnes and wrote for his online magazine *Street Carnage*. If you know anything about the right-wing extremist movement in the US, you'll be familiar with McInnes; he is a powerful voice in a cacophony of right-wing radicals. Initially a comedian, he cofounded the magazine *Vice*, as well as *Street Carnage*, before following a darker path. McInnes went from irony and biting humor to being a hipster who then gradually moved to the right.

Since then, middle-aged McInnes's activities have been anything but funny: he founded the Proud Boys, a group that has led violent, racist riots in the United States and has acted as if they were Donald Trump's unofficial protectors.

"When the alt-right attached itself to Donald Trump, it unleashed a kind of uncontrolled madness. The paranoia and conspiracy theories took over, and things got really serious," says Josephsen now.

In the darkest corners of the internet, Josephsen sees worsening stress and increasingly advanced conspiracy theories. He sees Norwegian teenagers involved with neo-Nazi groups that glorify violence; teenage boys posting hateful rhetoric, racist memes, and videos showing violence and Nazi propaganda. And they are getting younger, some of them are children. Researchers and all those who have been following this development are worried.

"Many of them come from the gaming scene but belong to a very special group of aggressive and frustrated gamers who don't understand social interplay," Josephsen says. "What you learn from gaming too much is that you get a reward when you complete a mission adequately. But that kind of logic doesn't work in the rest of the world. Gaming is a way of dealing with the world without actually dealing with it. And the conspiracy theory about Q shows us that gamer logic has moved into the real world."

In October 2017, a mysterious online profile named "Q" began posting cryptic messages on the websites Reddit and 4chan. Q's followers, called QAnon, include millions of Americans and many Norwegians too. Q's conspiracy theories develop according to a kind of gamer logic: those who have followed Q's mysterious messages develop conspiracy theories together. This has led to further conspiracy theories about corona vaccines, and another about how members of the US Democratic

Party were in a satanic-pedophile cult that met in the basement of a Washington, DC, pizza restaurant called Comet Ping Pong. QAnon was also the driving force when a large mob comprising a variety of groups stormed the Capitol on January 6, 2021.

"It was totally crazy watching all these people streaming in and the guy carrying Nancy Pelosi's lectern," Josephsen says. "I recognized almost every group and organization—those who'd been around from the start of the alt-right movement through the Trump presidency, from groypers, who are fanatical MAGA supporters, to accelerationists, who were only there to foment chaos. There they were!"

Josephsen wants to use the knowledge accumulated from his years in these toxic forums, hoping that by adequately explaining how people get duped by the mechanisms behind right-wing extremism, he can prevent people from wasting years of their lives on it.

"I think we can save those who are already in there, but it has to happen face-to-face. We have to talk to each individual," Josephsen observes.

Josephsen now uses his dark experiences to write essays and articles. He also occasionally works as a researcher for TV documentaries on right-wing extremism and is an expert commentator. Like others brave enough to have frequented the most disturbing corners of the internet, he tries to communicate what he has seen and reverse the trend. The author and journalist Øyvind Strømmen is doing the same thing. Strømmen has for years been following what's happening on the dark web, too, and describes how young men without a social network become easy prey for radical ideologies.

"The radicalization process often begins with personal circumstances," he says. "People that have weaker safety nets are consequently less resistant to extremism when they encounter it. They are deeply alienated, often have a dysfunctional family life, and drop out of school and work life. They are lonely."

Strømmen has written several books about the far-right movement and conspiracy theories and has spent years watching it develop. For him, it's about how the internet becomes an alternative community to the ones in the physical world that young people are often failures within. This community allows these people, most often young boys, to devise theories about the world and its mechanisms, as a group, in the safety of their own bedrooms.

"There is a strong social aspect to conspiracy theories," says Strømmen, author of the book *Giftpillen* [The poison pill] about these theories. "They are often all about who is outside and who is inside. Where the danger is, who the enemy is among us, and who the enemy is on the outside. And this all becomes an extensive interpretation of the world and what this world is simultaneously being threatened by."

There is a growing legion of mostly young men who reach out to like-minded right-wing extremists online. It is an increasingly strong movement, constantly gaining ground. Today, conspiracy theories have become mainstream, so much so that they are sometimes even repeated by members of the us Senate; during the Covid-19 pandemic, Norway, too, was rife with anti-vaccine conspiracy theories, based on preexisting theories about how "big pharma" and the authorities make secret plans together.

Conspiracy theories exploit our ability to think creatively, to see patterns and connections where there aren't any. Humans love to see patterns: we can see Jesus in a piece of burned toast if we want. And we like knowing what's "really" going on behind the scenes. But what's more dangerous is this: conspiracy theories divide the world into "them" and "us," where "they" always have a sinister agenda, and "we" are going to save the world. You become a crusader in a way you've perhaps never experienced before. You are suddenly important, a key player in exposing the malevolent forces normal people don't see. And conspiracy theories always end by blaming someone for the world going so badly, a group of people with evil plans: Jews, the authorities, feminists, Muslims, communists, political dissidents, homosexuals. Conspiracy theories lead down a dark path where you eventually conclude that the logical thing to do about these "malevolent forces" is remove them.

Long before anyone had heard of Anders Behring Breivik, Øyvind Strømmen examined the world of online Islamophobia, throughout which the much-discussed Eurabia theory is spread. The Eurabia theory describes the secret plan, which Muslim and Western authorities have prepared, for Muslims to take over the entire Western world by ensuring there are high rates of immigration and that Muslim immigrants have lots of children. It was a theory Breivik believed in. In 2007, Strømmen warned that there would be an anti-Muslim attack in Norway. But no one took him particularly seriously, and almost no effort was made to monitor the community that Anders Behring Breivik sought to be a part of—not until four years later, after his bomb exploded in the middle of Oslo. But what's unique here is that Breivik failed to gain acceptance from the people he admired. One of them, the far-right Norwegian blogger "Fjordman," showed no interest when Breivik contacted him. What emerges is a picture of a very lonely man who didn't even make friends among other right-wing extremists.

In his 2019 book *Hateland*, Daryl Johnson, a former analyst in the US Department of Homeland Security, writes about several high-profile cases involving solo terrorists and tries to find a common denominator. What these terrorists have in common, according to him, is something he calls the wall of frustration: people who commit tangible acts have typically attempted to become part of a radicalized online community—and failed. Only then have they acted: they justify their actions by claiming to be the only person who *really* understands what the ideology means, and the only person willing to do what it takes to be a "true" believer.

"I think this is a fairly plausible theory," Strømmen says. "Solo terrorists choose something that cannot dismiss them as brutally as humans can—a theory—and for these lone wolves the theory becomes more important than human life. It's possibly those who fail to even fit in socially online that become the scariest of all, i.e., who go on to become real terrorists."

So a terrorist is the loneliest of the lonely, someone who can't even find friends among people united by hatred and racism. You can safely say that right-wing extremism is an ideology of loneliness, since it essentially focuses on "us" and "them," and the "us" has so little room to spare, while those outside the inner circle will have to be pushed out, removed, and killed once a new Nazi world order is in place. Exclusion is a natural part of the ideology. The brutality of neo-Nazism doesn't scare off its followers, who all hope to qualify as members of this tenacious community. Because it has existed online for a long time. One of the first major far-right websites was founded before Google even came along—it is called Stormfront.

"This is a proper social-networking website," says Strømmen. "The people on far-right websites like these are not lonely—although they might appear lonely to bystanders. So it's important for us to create a more nuanced picture about the equation between the internet and loneliness. The internet is in itself a social medium, primarily driven by communication."

But there's something about the internet that exacerbates loneliness, that raises the conflict level. When I go on social media, the discord just screams at me: arguments between people who have perhaps never even met, writing to each other in all caps. In one debate, someone writes, "When I read comments like this, I'm almost sorry that my father's old gun has been plugged." People who speak out publicly about hot-button topics like feminism and immigration suffer harassment and threatening messages; some therefore choose to remain silent.

Social psychologist Jonathan Haidt is worried about how public discourse has changed. "Social scientists have identified at least three major forces that collectively bind together successful democracies: social capital (extensive social networks with high levels of trust), strong institutions, and shared stories. Social media has weakened all three," he wrote in a 2022 article for the *Atlantic*, ". . . social media amplifies political polarization; foments populism, especially right-wing populism; and is associated with the spread of misinformation."

Haidt's research shows that many of our political and moral beliefs are instinctive, and are only justified by logic and morality afterward, and these moral feelings are especially triggered by the algorithms of social media, feeding off our rage and fears. Haidt has constructed a range of absurd experiments that prove that we make moral judgments based on irrational gut feelings. For example, he asked large samples of people if burning a flag is okay provided no one sees it or is offended by it (most still say "not okay," even though the logical reasons to object no longer apply). He also thinks he can prove that our political views become more conservative if we are in the vicinity of a washbasin, perhaps because we then become more interested in cleanliness: ideas about cleanliness are strongly linked to conservative politics. What worries him is that he sees people building more and more of their identity on these instinctive opinions and using them to reject other people.

"The idea that we organize ourselves into tribes has gained a foothold, but the tribes of the past weren't constantly at war, as they are often portrayed," he says. "Real tribes were rarely at war, they shared their culture with other tribes, and it's at least the same story with humankind."

The mere fact that cooperating has been absolutely key to our survival disproves the brutal, social Darwinist understanding of early human life as being a continually hard struggle for survival.

"Human nature has given us the ability to erect walls between each other, but we are also very good at sharing and exploring," says Haidt. "Human nature is both of these things; we don't need to overcome human nature in order to avoid conflict."

Haidt is worried about how we nurture conflict and competition. In America, this notion is part and parcel of hard-nosed capitalism and the rationale behind New Public Management, both of which are shaping modern-day British and Norwegian society. Conflict and competition may well lead to economic growth (at the expense of the climate), but it also leads to ruthless exclusion and acute loneliness for those unable to assert themselves in the competition.

In their book *The Dawn of Everything*, anthropologist David Graeber and archaeologist David Wengrow also claim that conflicts between human groups are more unusual than you would think. "While human beings have always been capable of physically attacking one another (and it's difficult to find examples of societies where no one ever attacks anyone else, under any circumstances), there's no actual reason to assume that war has always existed," write Graeber and Wengrow.

But one source of conflict, according to the two scientists, is that we began comparing our own pack or group with other packs or groups. We then began exaggerating the differences we found, in order to reinforce our identity—what Freud calls "the narcissism of small differences": it is easier to compare yourself with people who are quite like yourself than with those who are totally different.

"Identity came to be seen as a value in itself, setting in motion processes of cultural conflicts," Graeber and Wengrow write.

The sociologist Norbert Elias describes how groups will unite against an external enemy when fearing for their own survival. The threats needn't be genuine, as long as they create unity. Common enemies are social glue. "From the earliest days, societies formed by human beings have been Janus-faced: inward pacification, outward threat," he notes.

So exclusion creates both unity and identity; right-wing extremists can be united in their contempt for feminists, the queer and the melanin-rich, Muslims and left-wing politics; and the more people they can hate, the stronger the unity, and the less they have to deal with their own outsiderness and loneliness. If you don't know yourself, you can find your place through violence. The Canadian professor Marshall McLuhan believed that violence builds identity: "Ordinary people find the need for violence as they lose their identities," he said. "It is only the threat to people's identity that makes them violent. Terrorists, hijackers—these are people minus identity."

Then we have people who voluntarily and without struggle exclude themselves from the community, who simply let go and give up. Who choose to withdraw from the community, such as hikikomori and incels.

Lasse Josephsen has been monitoring incel culture—the word *incel* being derived from "involuntary celibate." Comprising young men who believe they have no chance of having a sex life, it is a culture that is supported by theories about "alpha and beta males" and is spread by "manosphere" prophets such as Andrew Tate. In this worldview, the alphas get all the women. The betas don't get any. And the associated misogyny and aggression is something Josephsen has seen increase online.

"They're looking for community but aren't really part of any online community either," Josephsen says. "If one of them out there says he wants to kill himself, the others will happily suggest different ways for him to do it."

Incel culture's view of women is pretty dark. Believing that women are only interested in genetic winners, incels define themselves as genetic losers, betas, those who are already outcasts and lonely, based on something as inescapable as the genetic material they were born with. This allows them to explain their involuntary celibacy as being something almost predetermined. Many of them attempt to change their physical appearance, molding themselves according to ultra-masculine ideals. Eating disorders and body dysmorphia are already increasing radically among all young boys and men, many of them wanting to become more muscular and masculine. But body dysmorphia has a very special place among incels, who put a lot of time and effort into their appearance, and whose aggression is directed at women who reject men because of their narrow shoulders or weak jawline, or at the system that is to blame for the lack of sex and closeness, love and family in their lives. They feel excluded and overlooked. It is a movement driven by unadulterated loneliness.

"They don't realize that the loneliness they're experiencing is mainly self-inflicted, that they need help, and that it would be much better for them if they just dared to step out into the real world," says Lasse Josephsen, who is writing a book specifically about incels. "They vehemently believe in the importance of having the right look and body. They haven't realized that being charismatic can make up for whatever physical shortcoming you have."

For his part, Øyvind Strømmen is relieved he didn't find an incel community when he was a teenager and wanting a girlfriend himself. "I've thought about how glad I am that there was no online community like this when I was young, somewhere that gave me an 'explanation' for how things really were," Strømmen points out. "Because it's quite normal for young people to feel inadequate when it comes to sex, and when you're feeling most sorry for yourself and stumble upon a community that tells you that the reason you can't get a girlfriend is because we live in a society that oppresses men, you then have an explanation for why things aren't going the way you planned. It's a community where you find a lot of misfits, which is perhaps a far more logical reason for why they haven't found a girlfriend."

In Norway, some young boys and men have withdrawn from society altogether, vanishing into the internet behind the bedroom door. They, too, are becoming hikikomori, part of the aforementioned phenomenon that was first recorded in Japan.

"I noticed it when I began seeing fewer and fewer teenagers on the streets," says social worker Lars Aarvåg-Amundsen. "They disappeared. And parents, worried about their children, began contacting us."

Aarvåg-Amundsen has built a community for lonely teenagers in the Norwegian city of Bærum, offering meeting places such as Spillhuset, a gaming center run by the teenagers themselves. The community's main target group is aged fifteen to twenty-five, many of whom have dropped out of school. Most of them are boys. And a common denominator is loneliness.

"We don't want to make lots of diagnoses," Aarvåg-Amundsen says. "We try to focus as much on the individual reasons for these youths becoming lonely, as we do on the society around them. One of the boys that comes here had been on his own, indoors, for three years. He became more outgoing after a while, but was then diagnosed with autism spectrum disorder, and that somehow gave him a reason to withdraw again. Trying to understand why these young people don't feel like part of society is far more important than giving them diagnoses."

Aarvåg-Amundsen found that there was no single reason for teenagers withdrawing from the society around them; there were many. Family conflicts could be one reason, problems at school another; it could be linked to moving and rootlessness or bullying; there were lots of serious incidents at the heart of each case.

"When we began looking into the cases, we discovered that these teenagers had very good reasons for withdrawing," says Aarvåg-Amundsen.

But for Lasse Josephsen, it's obvious that young boys are particularly vulnerable. They are easily drawn into online radicalization because they are lonely. So what is it with young boys and men? Why do they drop out, why do they become so lonely? They become the lost children; they become Peter Pan. They become Tyler Durden from *Fight Club*, the story of a deeply alienated and lonely insurance agent who starts a bloody bare-knuckle boxing club to help himself feel more alive and important. In the film, Durden ends up carrying out a terrorist attack. Today, the '90s film based on the book by Chuck Palahniuk seems strangely visionary: violence and extremism as a kind of response to being unable to find a meaningful place in society. This sense of alienation seems to be a problem, especially for boys and men in Western society.

More than 30 percent of boys drop out of high school, with all the negative consequences it has on their careers. Young boys and men are near the top of the suicide statistics, and the numbers for this group are only increasing. Incel culture is almost exclusively reserved for boys, and those who become radicalized are mainly young men. Young men are gamers; they become hikikomori. They become school shooters; they are rapists. Older men become lonely when their wives die or divorce them, because they never acquired their own circle of friends. Men kill their wives, not the other way around, and violence carried out by men against women continues to be widespread. There has to be something about the male role that doesn't really work, that has gone off course. Most men find their place in the world nevertheless. But the male role is a lonely one in itself: you cannot show weakness, or ask for help, or cry, or talk about your feelings. The need for boys to be strong and able to fend for themselves is present from a very young age.

A boy must be an adventurer and a conqueror. And that is a doorway to loneliness.

"How is it possible for men of my generation to develop such xenophobic and anti-feminist views," asked researcher and author Anne Bitsch in a series of articles in the newspaper *Morgenbladet* in 2021 that marked the tenth anniversary of the July 22 attacks. Anders Behring Breivik was a typical outsider. When he was arrested on Utøya, he claimed to be a member of a secret order called the Knights Templar. This, however, turned out to be a figment of his imagination. He was fundamentally lonely.

"Maybe that's what happened to Anders Behring Breivik," Bitsch writes, "he never found his place in any community and therefore had to invent one of his own; he couldn't cope with his own insignificance and so chose to write himself into history in a way that would prevent anyone overlooking him again? In his bedroom at his mom's apartment, he became lost in contemplation of a pending apocalypse and spent his time gaming and reading Islamophobic blogs. Maybe it was the lack of contact with normal people, and constant exposure to rhetoric that values distance more than community, that made him crack."

The Norwegian documentary filmmaker and activist Deeyah Khan is seeing the same problem: Why is it specifically boys and men that vanish into these online communities? Why do boys and men become violent extremists? Khan has used her skills as a documentarian to understand more.

"Nobody talks about it," she says. "But today's men live according to the demands of a toxic male role. Which is so limiting, there's so little space. It's the dehumanization of men. When we create such confined spaces for each other, we *dehumanize* each other."

In her award-winning documentaries, Kahn has presented extremists of all political stripes, including both Islamic and right-wing fundamentalists, and most of those she meets in these communities are men. One of the big problems, she believes, is the notion of what men should be, what kind of agency men have. Another problem is loneliness, which runs through everything.

"I picked up a camera, they picked up a gun," she says. "That's the difference between us! We must learn to get a grip on loneliness before it is exploited politically. There are no monsters out there, but it's easier to think so."

Khan has won a prestigious Emmy Award; she has created a magazine for Muslim women and is UNESCO's Goodwill Ambassador for artistic freedom and creativity. As a high-profile feminist and human rights activist, despite the darkness she has seen, Khan still believes there are more similarities than differences between her and those she profiles in her films: male right-wing extremists and jihadis, murderers and anti-abortionists. And she has experienced acute loneliness herself. For her, fleeing the racism and death threats she experienced as a female Muslim musician in Norway felt like falling out of society completely. After moving to Britain, without having a single friend or acquaintance there to help, she found that she no longer wanted to continue with music. She was absolutely lost. She didn't know which direction to choose in life anymore and found herself totally outside the society she was now living in. So she very much understands the sense of floating in space that extremists of all kinds have felt—extremists who have often been the subject of her investigations. Her current documentary is about a murderer.

"As a rule, I think we could stop it, we could stop the violence and the extremism," Khan says. "It boils down to who was there when you needed someone, who offered you a community. Because people always need other people. A terrorist doesn't exist in a vacuum. ISIS fighters, Breivik, all the lonely men who've experienced trauma, hopelessness, and helplessness—they want to be heard. They may have asked for help or signaled that they need help in the past and will have been surrounded by teachers, extended families, local communities who haven't seen them. We don't see them until there's a terrorist attack. And we then prove them right, because it's only then that we listen. It makes me so angry!"

She does not believe that combating loneliness and extremism should solely be the job of state institutions. It cannot be exclusively part of a public strategy. The small acts of kindness we do for each other face-to-face cannot be enacted at the government level. She believes it is everyone's responsibility to combat loneliness. And this responsibility is perhaps especially clear during school years. Being good at noticing children who are suffering violence and abuse is about humanity and care, and about giving love to youngsters who are feeling lost.

Because it shouldn't all be left up to professionals. Harvard professor of social medicine Vikram Patel is a passionate supporter of grassroots work in local communities. Organizations like Mental Helse work to ensure that all children get health-promoting and preventive education about mental health at school. If we are all better informed about what depression, violence, PTSD, and loneliness entail, we can become a richer and more generous society. Because a society where we are closely woven together—from the moment we encounter the provisions of the welfare state to the open arms of neighbors and friends, where we are surrounded by love from the start—is a place that right-wing extremism and violent ideologies become a less interesting alternative.

"The bad news is that there is no single measure that can solve the problem of loneliness," Khan says. "The good news is that we can draw individuals away from extremism, one by one, using human warmth. I think life's too short to not behave deeply human, for the sake of others and, perhaps mainly, for ourselves—I try to understand in order to find my own humanity. I'm always looking for that fundamental thing we share. Behind acts of violence and terrorism, there is always, without exception, someone who is fundamentally lonely. It's the reason I work as an artist."

Khan now lives in the US, where the welfare system is very poor and public trust in the state is low. It is a recipe for loneliness. A recent study reported increased loneliness among American teens over the past decade and connected it to the use of social media. But for prosperous little Norway, where we have good welfare schemes and a lot of trust in the state, the mystery remains. During the pandemic, as mentioned, the number of lonely Norwegians between the ages of sixteen and nineteen increased sharply, to 70 percent.

"Norway is a very well-functioning society; we have lots of good welfare schemes in place," Khan observes. "But it's about more than that. We have to be able to care for each other."

Khan's message of "radical empathy" has reached people far and wide. When she meets extremists, she changes them. While filming *White Right: Meeting the Enemy*, she spoke to men with racial prejudices against people just like her: a young Muslim woman. However, many of these men later renounced the far-right movement, and it was Khan they called for help and support when cutting ties with the violent organizations that had once offered them a community.

Back in 2011, in the square outside Oslo City Hall, I felt a large and powerful sense of community that would soon evaporate. I had momentarily felt the electric energy of that massive crowd spontaneously bursting into song. It was nice to suddenly feel part of something bigger, to forget myself and become like everyone else, surrender myself. I found a similar community in Moominvalley, when reading my daughter's bedtime story: a community called the Hattifatteners, who are totally identical and group and cluster when storm clouds gather above. It is hard to tell if this grouping is because of the impending lightning strike, or if the grouping is what's causing the storm.

"Moominpappa felt an irresistible desire to do as the Hattifatteners did: to sway back and forth, to sway and howl and rustle." But when the thunder abates, he wakes up and returns to being his normal self: "He looked at the Hattifatteners, and with electric simplicity he understood it all. He grasped that only a great thunderstorm could put some life in Hattifatteners. They were heavily charged but hopelessly locked up. They didn't feel, they didn't think—they could only seek. Only in the presence of electricity were they able to live at last, strongly and with great and intense feelings."

Vanishing into a community might be nice, but it's also dangerous. Community can offer strength that can be used positively and negatively; nobody wants the sound of marching Nazi boots to return to our streets. We do need groups, of course, just not such regimented ones. Moominpappa, who is essentially quite a lonely figure, an orphan and an independent adventurer, tears himself free: "He was himself once again, he had his own thoughts about things, and he longed to be home."

That night, I walked home from City Hall Square knowing that I belonged to many different groups and had my own opinions on many different things. Good communities need to have room for diversity. This is the paradox of humanity: we want to be part of large communities, but we also want to be individual and special. And both of these needs have to be somehow balanced within good, democratic, and inclusive communities.

The future Deeyah Khan dreams of is rather like Moominvalley, with its own Little My and Snufkin and Too-Ticky and Sniff and Snorkmaiden, Moominpappa, Ninny, and Groke. Everyone in this future is uniquely strange, and busy with their own projects, and sometimes they feel a bit lost and lonely, not fully understood. They don't march in lockstep, nor are they identical like the Hattifatteners. But this Moominvalley is otherwise a small community made up of peculiar individuals—Khan and former extremists—who gather around the fireplace in Moominhouse and share their stories.

"I think loneliness is a good basis for a real community," Khan says. "We are so many, and we are so alone, and our recognition of this fact is what brings us all together and provides the societal glue.

"It means that people like me can have a natural reason to share a table with renegade neo-Nazis. So many people have experienced violence and loneliness! We just have to find our people! And they're not who you think they are, they don't look the way you'd expect—but they're out there. That's what I want to give my daughter, that will be my legacy to her: a large and strange community."

6

Why it is so difficult to trust each other

WHEN ASKED ABOUT loneliness in 1987, Charles Bukowski told *Interview* magazine that the loneliest thing he could think of was crowds. "It's being at a party, or at a stadium full of people cheering for something, that I might feel loneliness."

The poverty Bukowski grew up in, along with a face scarred by severe acne, meant he always felt visible in a negative way, and outside the community. He also grew up with violent and unpredictable parents who would beat their son: it was a very basic case of neglect from what should have been his most important flock. Of course, crowds aren't necessarily safe anyway, as we already know from the previous chapter. But this insecurity can be especially strong if you've already lost your basic trust in people as a child. If you can't trust the people in your immediate vicinity, you'll struggle to overcome loneliness. The bridge, from the mainland to your island, will remain impassable.

The biggest cause of loneliness in a society, according to many researchers, is lack of trust; as I mentioned in the previous chapter, trust and loneliness are connected. Where there is no confidence in society at large, where institutions are rife with corruption and nepotism, or where there's no network looking after the citizen's need for protection, knowledge, health care, and schooling, we find more loneliness than in societies with highly dependable and well-functioning institutions. Living in an insecure society, where values like the rule of law and freedom of expression go unprotected, will make us lonely; we would then know that our voices and stories of abuse and violence will not be heard. Paradoxically, the connection between trust and loneliness does not apply to Japan—where there is both a generally high level of trust in society *and* a large and increasing amount of loneliness. There are, in general, quite a few paradoxes when I look at the connection between trust and loneliness, such as the strange fact that Italy—with its extremely close families and a collectivist culture, its shared meals, celebrations and opera in the summer, and Grandma's spaghetti and minestrone in a warm home in the winter—should rank so high on the loneliness statistics, simply because Italians have so little trust in society. There is, of course, a fair amount of corruption and organized crime in Italy; and the more unclearly the hierarchies are organized, the more difficult it is for citizens to be heard and understood. And the smaller and more unimportant and unprotected you feel, the more likely you will consider yourself to be in the outer circle of loneliness. You become invisible and powerless; you become lonely. This could be why researchers believe there is a general connection between perceived trust at the population level and the loneliness reported at the individual level. Because a society that so easily allows people to slip through the net will fail to protect us, and in turn put us at constant risk of ostracism and violence. And our fear of this happening makes us withdraw from each other. Trust is about security, openness, and joy, and having a deep connection with other people. Without trust, it is very hard to create well-functioning societies; they get torn apart by conspiracies and poisonous rumors. When people don't trust each other, they work against each other.

Lack of security and belonging is the reason for the global crisis of loneliness, says author Noreena Hertz, who is described as one of the leading experts on economic globalization.

"Even before the coronavirus hit, we were part of a global crisis of loneliness," she says, "as I crisscrossed the globe researching my new book *The Lonely Century,* I was struck by how huge the range of people profoundly affected by feelings of disconnection and isolation was."

Loneliness has increased, and Hertz believes that some of the main reasons for this are capitalism, the internet, and the way our cities are organized. A society that doesn't have space and time to cultivate small acts of kindness is a lonelier society, she says; a society driven by profit and efficiency is not typified by connection and belonging.

One striking experiment looked specifically into this: in 2013, sociologists Gillian Sandstrom and Elizabeth Dunn discovered how just thirty seconds of friendly conversation with the barista at a coffee shop made their subjects feel happier and gave them a stronger sense of belonging for the rest of the day than the control group who had to be efficient and avoid small talk while in the coffee shop. These micro-interactions can bring city dwellers a little closer together and create trust; even a fake smile will affect our sense of togetherness and joy. But now smiling and eye contact are becoming a rarity, and during the pandemic all these small meeting points almost disappeared entirely.

"At work, in open-plan offices, people counterintuitively are more likely to communicate via email, rather than talk face-to-face," Hertz points out, referring to workplace research.

She believes that we communicate less openly and directly because of stress and the demands placed on us for efficiency, and that these factors contribute to the degradation of real contact and our sense of belonging. And being connected to the society around us is good both for our health and for a functioning democracy.

Vivek Murthy, the surgeon general of the United States, points out that all the little acts of kindness that surround you, from a few extra words with the guy in the coffee shop to small talk with other parents beside the football pitch, help you feel like you are part of something bigger.

"Kindness, appreciation, and generosity are as essential in brief interactions with strangers as they are in closer friendships," Murthy writes. "These exchanges take only seconds, but they can create a meaningful sense of connection," he points out.

But it's possible that these small altruistic gestures alone cannot create trust. Governments, such as the one in the United States, that don't provide free health care to their citizens do not inspire trust; because you know that, without the right health insurance, getting an injury or a serious form of cancer can be disastrous. If you know that by arriving fifteen minutes late to your appointment at the welfare office you will have your benefits suspended, as is the rule in Great Britain, you will also be fully aware that should a crisis hit, you will be entirely on your own. Trust is built up from the ground up, from the citizens themselves, but an economic and political system that doesn't foster confidence and care will create lonely citizens. Capitalism and New Public Management prescribes that everything we do, including how the health care sector is run, should be measured, weighed, and sold. In such a world, we are objects, producers, pieces, and this failure to understand who we are and the resources we possess makes us lonely. It creates lonely citizens, because it makes being a success more important than belonging to a herd.

The pace of life, meanwhile, is just accelerating; workplaces are becoming more streamlined; and there are many people living alone. Ironically, the more civilized and well-functioning towns and cities become, the lonelier they feel. Societies founded on ideologies and belief systems such as neoliberalism and capitalism—which claim that we are entirely self-sufficient individuals who are responsible for our own successes (and thus our failures too)—primarily believe that everyone should fend for themselves. But if you make your own luck, you are at the mercy of yourself. Over 50 percent of London's and New York's inhabitants report being lonely.

A new report published by the British think tank Onward in July 2021 states that there is an unparalleled epidemic of loneliness in that country. In just ten years, the number of British under-thirty-fives who claim to have only one or even no close friend has tripled. Those with four or more friends have fallen from 64 percent ten years ago to 40 percent. Millennials don't participate in group activities or their neighborhoods, and only 30 percent of people under thirty-five trust their peers. In 1959, when asked if they could trust each other, 56 percent of teenagers answered yes, while only 30 percent of young Britons say the same today. This confirms the trend in the United States, as documented in Robert D. Putnam's classic *Bowling Alone*, a book that describes how, over a very short period, Americans have stopped participating in local communities and associations. It becomes a self-reinforcing, vicious spiral of withdrawal and loneliness. Onward proposes a range of economic measures that should increase confidence among young adults, such as rewarding community service with student loan repayments; allowing local communities to make disused premises into sports clubs and welfare facilities; and sponsoring affordable youth housing to give young adults an earlier start in the housing market. That Onward director Will Tanner mentions the economy is no coincidence. Poverty is connected to loneliness; it is about shame. It also makes life far more difficult, including everything that might counteract loneliness. You don't have the energy for community when you're struggling to pay the bills.

"Young people are suffering an epidemic of loneliness that, if left unattended, will erode the glue that holds our society together. After decades of community decline and 15 months of rolling lockdowns, young people have fewer friends, trust people less, and are more alienated from their communities than ever before. And it is getting worse with every generation," says Tanner in an interview. "If the pandemic has taught us anything, it is that human connection and local place deserve a much greater place in our political debate than they have enjoyed in the past," he continues.

"Qualitative research for this paper reveals that young people are not detached from their communities out of choice, but through lack of opportunity, security and time," Onward's press release says.

That's why the social glue is so important: you're better off when you experience lots of small moments of connection than you are if everyone around you is a potential threat. And when society doesn't offer security, it creates a negative spiral where citizens start to mistrust each other. In the aftermath of the pandemic, economist and former member of parliament Marianne Marthinsen felt that the Norwegian welfare state was at a breaking point, with financial decline and a climate crisis on the horizon: trust is like a sweater with a loose thread, and if you pull on the thread, it can all suddenly unravel.

"I find that my own trust in society is dwindling," she writes. "So far it's about relatively small things. I don't fully trust that I'll manage to get what I need from the shop, that I'll get from A to B when I need to, that the pharmacy has enough of the medicine I've been prescribed."

In his book *Together*, Dr. Vivek Murthy points out how increasing numbers of immigrants and migrant workers, internally displaced persons and refugees carry stories of intense loneliness. In such circumstances, there are many obstacles to finding friends and networks: you don't understand the language, you're looked down on and subjected to racism in your new country, you don't connect with the network of little communities that keep the rest of us safe. Researchers Alberto Alesina and Eliana La Ferrara have found that people who, for example, have experienced racism or other forms of trauma are less trusting of society. Not knowing who to trust is one of the loneliest things there is. It means that no one is your friend. Because without trust everything does become if not impossible, then extremely *difficult*: How can you buy something from a person if you don't trust they will give you the eggs and milk you paid for? How can you send your child to a school if you don't trust that the school will take good care of them? Who should you vote for if you think that politicians just feather their own nests? If you don't trust anyone around you, you won't make friends and you won't find a partner. Citizens who are unable to trust are unable to build a society. Mistrust is like poison in the tap water: it permeates everything.

Trust is the very glue that holds society together.

When we are struck by an act of terrorism, the degree to which our society is built on trust and cooperation becomes even clearer. It's the everyday things that go unseen, that are perhaps not appreciated enough. I can't help but think that a society that makes us see neighbors and colleagues and class-mates as competitors and adversaries, not as potential friends and collaborators, will make us less connected to each other. Yes, small, kind actions are important, but if we don't endeavor to build a society that is essentially about looking after and taking care of individuals, then our trust in each other will perhaps crumble, and we will lay the foundation for a loneliness pan-demic. If you know you are dispensable and not wanted enough to be cared for when you are at your lowest ebb, you will natu-rally fear ostracism and the subsequent financial and personal crisis. That is loneliness. I guess I'm just trying to say, in a very simple and naive way, what Noreena Hertz writes in her book *The Lonely Century*:

> For democracy to function well—by which I mean to fairly reconcile the interests of different groups while ensuring all citizens' needs and grievances are heard—two sets of ties need to be strong: those that connect the state with the cit-izen and those that connect citizens to each other. When these bonds of connectivity break down; when people feel they can't trust or rely upon each other and are disconnected, whether emotionally, economically, socially, or culturally; when people don't believe the state is looking out for them

and feel marginalized or abandoned, not only does society fracture and polarize but people lose faith in politics itself.

In lonely people, the part of the brain most closely associated with empathy declines. The body and brain are tuned so that if you are often lonely, you'll be more focused on the possible threats or dangers from other people than open to their perspectives and wanting to understand them. Because if you're worried about being expelled from the pack, you won't have the energy to show empathy; you will primarily be looking for safety.

It's only while working on this book that I've realized how much my own distrust has colored my relationships and experiences. And the most painful thing about writing a book on loneliness is perhaps that I've come face-to-face with the Groke. She is me. But in a society without trust, any one of us can easily become like the Groke: the loneliest person in Moominvalley, walking in silence while spreading an icy chill. The Groke never opens up to anyone, even though she is clearly sad and lonely. She cannot love or express even the smallest thing she wants. And although the Groke needs the Moomins, all she can do is stand by the house, seemingly uninterested in those inside, while just staring at the storm lamp outside. The Groke trusts nobody.

"It's called epistemic trust," Oslo-based psychologist Peder Kjøs says. "It means having a basic willingness to trust that the world is a good place and that people can be trusted—an expectation that the world is good. If you're not brought up with this expectation, you have to learn it yourself. And that is very difficult."

Working with trust is a continuous process, he believes, a kind of balancing act. Sometimes too trusting, sometimes not enough: it's constantly changing.

"It's something we're all continually adjusting," Kjøs says.

Trust is a kind of dance with other people whom we recognize from our connections with them, from our understanding of them, the eye contact and the laughter, all things I've written about in previous chapters. It is not a static feeling, but something that's constantly evolving. We are continually assessing whether we can trust other people. If they turn out to be unreliable, most of us manage to adjust accordingly without too much damage being done. But if you already lack a basic sense of security, any small breaches of trust will be hard to cope with.

"The Groke shuffled a little nearer," writes Tove Jansson in a crucial scene in the Moomin literature. "She stared into the lamp and softly shook her big, clumsy head. A freezing white mist hung round her feet as she started to glide towards the light, an enormous, lonely grey shadow. The windows rattled a little as if there were distant thunder, and the whole garden seemed to be holding its breath."

It is here that I realize that something has made the Groke turn cold.

"Mamma," whispered Moomintroll. "What happened to her to make her like that?"

"Who?"

"The Groke. Did somebody do something to her to make her so awful?"

"No one knows," said Moominmamma, drawing her tail out of the water. "It was probably because nobody did anything at all. Nobody bothered about her, I mean."

The Groke is loneliness personified, a dark and cold being who is unable to leave those inside the house in peace but also unable to embrace their love. And her story begins with nobody caring about her. It begins where all stories of loneliness begin.

In recent years, a lot more research has gone into finding out what it takes for us to bond with and trust each other. Having no trust in the people around you lays the groundwork for a lonely life. And daring to trust others is linked to the very first connection we make. This initial moment of trust is where it begins, our lifelong entanglement with other people that is so determinative in whether we feel lonely or not. The late German-American psychologist Erik Erikson outlined the eight most important psychosocial development stages in a person's life, the first one being trust. If trust doesn't form early, during the first eighteen months, everything in the child's life will be built on mistrust. It will affect all subsequent relationships. Can the innocent child trust that it will be attended to and cared for, that it will be given care when required? If not, it's like building a house on quicksand: trust is the most fundamental thing in life. And if you profoundly mistrust the world, trusting anyone in it, even trusting yourself and your own judgments, will be difficult.

Researchers believe they can see this right from the start— how the interaction between a baby and its caregiver sets strong guidelines for the rest of the child's life. It might sound extreme; nevertheless, a well-known study, conducted in Minnesota of two hundred mothers and their children, followed the children until they became adults while looking for clear attachment patterns.

"For 30 years we have been wrestling with a key question in developmental psychology; namely, do individual patterns of adaptation emerge in a coherent manner, step-by-step, beginning in infancy?" wrote the research group when summarizing their work.

The study began in the mid-'70s and wasn't completed and ready for publication until 2005. This enormous project, led by professor of child psychology L. Alan Sroufe, would have a huge impact on our understanding of what attachment is, what it does to us, and what effect it has on the relationships we have in later life with people other than our parents: what kind of partner we choose, how easily we make friends, and what kind of parents we become ourselves.

The researchers behind the study were building on the theories of John Bowlby, the pioneering child psychologist who launched the widely used term *separation anxiety*, which is really just another term for loneliness. Bowlby explains separation anxiety from the perspective of evolutionary psychology: the child naturally becomes afraid of being abandoned by the person who is essential to its survival; *it becomes afraid of being abandoned by its most important flock, with a potentially fatal outcome*—which is, after all, the very definition of loneliness, according to researcher John Cacioppo.

After the Second World War, Bowlby was commissioned by the World Health Organization to research orphans, and in 1951 he was therefore able to launch his theory about the crucial importance of parent-child attachment for positive child development. In 1970, Bowlby's younger research colleague Mary Ainsworth developed his theories further and outlined three types of attachment. To understand the different ways in which a child between nine and eighteen months of age bonds with its parents, Ainsworth developed the "Strange Situation," an experiment still used today by child protection services and child psychologists. The experiment involves a mother leaving her child alone in a room and then returning. Later, a stranger enters the room where the mother and child are, and the mother once again leaves, only this time the stranger is left alone with the child. In the end, the strange situation isn't that strange, because the mother eventually returns, and the stranger leaves the room. Nevertheless, it allows the researchers to observe the child in the eight different situations that arise, either with the mother or with the stranger or both, to see how the child handles each situation.

A child with a secure attachment style will happily greet its mother when she eventually returns and carry on playing, while a child with an insecure attachment style will turn away from its mother, or even resist cuddling and closeness, when she finally enters the room. The researchers also observe how the child copes with its mother not being there—how scared, happy, and exploratory a child is in this strange setting. Most of the children Mary Ainsworth examined, roughly 70 percent of them, had a secure attachment style. But about 15 percent both feared the stranger and rejected the mother when she returned to the room, which Ainsworth described as ambivalent attachment. Some children displayed no signs of fear when the mother left the room, or when the stranger entered the room, and were, additionally, quite indifferent to the mother when she returned. Ainsworth called this avoidant attachment, and it can be interpreted as a result of neglect. The child's indifference is a response to the fact that there is no genuinely secure connection between mother and child.

Ainsworth's student Mary Main, who became a researcher herself, added a fourth category to the existing three: disorganized attachment. In the Strange Situation, a child with disorganized attachment will behave in a contradictory way, show signs of fear or stress, freeze or smile for no apparent reason, but at the same time look happy when the mother returns.

This child sends out many signals simultaneously, because it is both afraid of and emotionally attached to an unpredictable and violent caregiver, or a person who is sexually abusive. These four attachment patterns were what the researchers behind the Minnesota study were looking for and what they attempted to follow into adulthood.

"It is an organizing core in development that is always integrated with later experience and never lost," the research group behind the Minnesota study concluded.

"Attachment experiences remain, even in this complex view, vital in the formation of the person," they write in their summary, which considers all the possible variables before looking at how the attachment patterns repeated themselves through adolescence and into adulthood.

The researchers found that these four patterns had an astonishing effect on how the children's lives played out. It is easy to believe that the researchers simply found what they had wanted to find, but they had prepared for this by observing each child thoroughly, in many different situations—and not just directly, but also by interviewing teachers and others who had contact with the child. They even interviewed the mothers before the children were born, to find out more about what expectations they had for motherhood, and carried out thirteen observations of the child in the first thirty months. What they looked for were indications of violence or rejection in the children's behavior; they looked for the consequences of neglect.

The researchers chose mothers who were fairly poor and lived in the inner city, because they knew that poverty increases stress levels; uncertainty and instability will worsen the situation for both mother and child. Research into violence against children shows that violence and neglect occur in all walks of life, but become more serious with increased economic uncertainty and poverty. So the researchers behind the Minnesota study expected to find more children with avoidant or ambivalent attachment styles in this group than they would find among middle-class families—and it turned out to be true. They found almost double the amount. They also found that as many as 30 percent of the children from the poor areas of Minnesota had a disorganized attachment style, which is the most serious and which causes a number of different personality disorders.

"According to Bowlby, malevolent attachment experiences, especially a contradiction between one's own experiences and what one is told has been the case, can lead to a constellation of factors, including 'chronic distrust of people, inhibition of their curiosity, distrust of their own senses, and a tendency to find everything unreal,'" Professor Sroufe and the research group write. "These are hallmarks of major personality disorders, including borderline personality. It is our position that serious personality disorders, on those rare occasions when they do occur, will be the legacy of disorganized attachment, at times in conjunction with avoidant attachment (and thus a combination of alienation and a tendency toward dissociation)."

Dissociative identity disorder and what's called borderline personality disorder are linked to fundamental mistrust, according to the researchers: you trust nothing and thus either become totally disconnected from your surroundings or have a constantly ambivalent relationship with them, including the subsequent mood swings and problems with regulating your emotions. This leads to acute loneliness and to psychological problems that make building even the most basic trust in others difficult.

Interestingly, throughout the Minnesota study's thirty-year duration, the researchers observed how the children with a secure attachment style became the center of their group of friends. They assumed leadership positions, easily navigated complex social situations, became teachers' favorites, and, in their adult lives, had good and close relationships with partners and children. The children with an insecure attachment style were seen as clingy, needy, noisy, or would quietly fade into the back of the classroom to avoid being noticed. They were suspicious and showed less empathy toward other people. The connection they experienced as children shaped them, to a greater or lesser degree, for the rest of their lives. This initial tone, which was set permanently, would reverberate through everything that followed. It's very hard to develop secure attachment in later life when you have grown up with a disorganized attachment style. Basic security is very complicated to build up afterward.

What attachment psychologists examine are tiny things, like how often a child turns to its caregiver, what the interaction between them is like, whether a child looks for adults other than its parents. The foundation of good mental health is laid during the interaction with the caregivers. A child who doesn't trust its carers and cannot interact well with them, who sees the surrounding adults as unstable and unpredictable, will not feel like a participant in its own life. A child like this will feel lost.

The researcher Daniel Stern believes that this can lead the child to create a so-called false self, a facade, and in many ways to lose touch with its true needs and desires, because it is unable to express them and have them responded to emotionally. Remember that young children are extremely emotional; it's only later that they're able to regulate and hide their feelings, with an adult's help. A child who isn't cared for and responded to can become numb and "dead" in relation to its surroundings; it will construct a "rejected self" detached from its inner strengths and emotions. This in turn could be one explanation for narcissistic personality disorder. Shame and lies contribute to this disorder, which is characterized by an unstable and fragile self that the sufferer has to cover up and beautify for fear of not being accepted by others. This diagnosis was among several given to Anders Behring Breivik (antisocial and narcissistic personality disorder). It is also the diagnosis that many somewhat overzealous American social analysts believe is affecting a whole generation of young people, those who have grown up with social media: it gives them superficial confirmation about purely external traits. In Stern's sense, this perhaps reinforces the sense of being lost. But that sense of being lost really begins with the most important group in your life, the family you grow up in: If you don't find security there, how can you ever find it somewhere else?

"Lots of children can find what they don't get at home somewhere else. At school, for example, or in kindergarten from teachers and other adults," psychologists Stig Torsteinson and Ida Brandtzæg point out.

One good person can be hugely important, just one person who sees and understands the child. Because children who are not cared for and understood at home will search for love and acceptance elsewhere. You need to find love somewhere if you want to survive. But what if you don't even have that one person? Who should you trust then?

"It was as though my entire childhood was a bomb going off in slow motion," says Marthe Bødtker, the daughter of a Norwegian Nazi. Secrets in a family can have the potential to explode.

Bødtker grew up with this same profound sense of insecurity, although as a child she assumed it was just how things were supposed to be. Her family was always on the move, with a new home every two or three years. But she adapted, as children do.

"Eventually, I gave up making friends," she says. "From the age of twelve, books were my friends, and I spent most of my time reading. Bringing someone home wasn't possible, because I was ashamed; I couldn't just tell people before they came that my father was an alcoholic and we had to pretend he wasn't. I was a lonely child."

Growing up, Bødtker at first idolized her father, and as he was often absent she would miss him enormously. It took awhile before she recognized that he was an alcoholic. He would stop the car and say he was going out to check something, then run to the back and start swigging from a bottle he pulled from the trunk. He would stop outside a pub, down a couple of beers, and come running back out to the car. He was always "working" in the carpentry shed, but it was important that she never walked in on him unannounced. Because it would then be real, he would then know that she'd seen him drinking, and this would strip him of his dignity. And if she caught him off guard, she didn't know how he might react. Would he hit her? It took even longer before she understood what it meant that he had been an active Nazi—a fact she learned more than fifty years ago.

"When I first heard about it, aged ten," she says, "I was slightly impressed! It felt a bit like having a pirate for a dad. Then it began to dawn on me what it actually meant. And later, I read his file in the National Archives and realized that he didn't just accidentally fall into Nazism. He chose it. He was fully aware of what he was getting into."

Bødtker describes her father as a kind of "Nazi Forrest Gump": he was always in the thick of it, at the center of the action, intentionally or not, and he took part in Operation

Barbarossa, the invasion of the Soviet Union in 1941, which killed around twenty-seven million people, mostly civilians— an invasion where Jews were often executed in the most brutal manner, before the industrial efficiency of the gas chambers. Six million people died during the offensive her father was part of, and in which he was eventually wounded himself, when he reached Dnipropetrovsk. He was also in Lithuania, where the persecution of Jews was equally grotesque. But despite being wounded three times, he returned afterward to the battle zone, fully aware of what was waiting for him. He was even in Berlin when the dark heart of Nazism fell to the Allies in 1945, but managed to flee north and narrowly escape Russia's invasion of Germany. He thus managed to return home to Norway, where he was arrested; throughout the war, he had also worked with recruitment and was a highly committed and well-known Nazi in his hometown. The family home contained objects that had been stolen from Jews who had either escaped or been sent to Auschwitz. After the war, the Bødtker family lived in constant fear of being discovered.

"For my parents, it was very much about having an overview of the social landscape we were in: Who was on which side? Our side, or the other side?" she says. "If there were too many people around us who knew who we were, we had to move. We were seeing the people through a particular lens; they weren't just people, they were either with you or against you, and it was through this filter that we always saw the world. We could tell from a surname if a family was on 'our' side or on the other. It was vigilance that spread to us children. Other people were a threat."

The other side was basically the winning side. And Marthe understood early on that she didn't belong there. The home she lived in was unsafe and unpredictable, and her trust in the out-side world was broken down too.

"It wasn't like I had a sense of community with Mom and Dad," she recalls. "It wasn't us against the world. I just picked up the shame and owned it."

Within the four walls of their home, they were against the world, but they weren't protecting each other. She was meant to associate only with family, and many of them were Nazis: her mother was fined and lost her civil rights, her father was in prison for several years after the war, and there were aunts, uncles, and her parents' friends...

"But why didn't you become a Nazi as well? Have you thought about that?" I ask.

"When I was eleven or twelve years old, I realized that my father, this person I'd adored, was not a kind man," she says. "It was suddenly obvious to me. Had he been kind and charm-ing, I might have allowed myself to be tempted. But it gradually became clearer how mean he was. He called my sister a whore, and said that I was going to be a whore, like my sister. It was a constant stream of malicious and disparaging comments. He also beat my older siblings, but I wonder if he'd become too old and alcoholic to beat me."

The malicious comments, the mocking, the shame, the secrets, the loneliness—they haven't gone away. To this day, Bødtker carries it with her, despite being over sixty years old.

"I remember failing an exam and thinking, *Now I'm everything he said I would be—a stupid failure of a girl.* I was someone he was ashamed of. I'll always be afraid, I think. And I still find trusting people difficult. I don't invite people to my house. Large crowds, like parties, feel scary; I always end up quickly sneaking out."

She doesn't even trust herself and her own memory, and lost

faith in her own judgment years ago. "I recount what happened to me over and over," she says, "because I often think: *Maybe I didn't experience this. Maybe nothing happened at all. Was it a dream, was it that bad?* I wonder if I'm mistaken."

No one in her parents' generation has ever been willing to talk about what happened or acknowledge what it did to their children. They pretended it was normal. She has lost years of her life: the stream of humiliation and fear of her drunken father has luckily been erased from her memory. And this is quite a normal way to feel; when experiencing extreme stress, you don't store memories in the usual way. But that doesn't mean you're not affected by them.

"I often think that I lived under a regime that denied me my puberty, you know, the time when your edges become more rounded, you get a little resistance and acceptance and become part of a group," she says. "That never happened to me. I had to make up for it in later life, and that's a bit strange, and it also reinforces the outsiderness. I'll very often become scared and withdraw; it takes a lot for me to feel safe, although it does happen sometimes. I've often thought that I'm a monster, and the child of a monster—that I'm *like that* and *born like that*, and that my life will never be good."

She now believes that life may never be entirely good, but it has become better than she could have imagined: she has friends, a job, and is surrounded by plenty of nice communities. From the outside, it's impossible to guess how she grew up and what she has experienced—how little trust she allowed herself to have in the people around her as a child. How much fear she had to live with.

This fear is something Martin Eia-Revheim experienced when he was growing up. Revheim, who founded the Oslo jazz venue Blå before going on to manage the Norwegian Radio

Orchestra, now manages a large multipurpose event center called Sentralen: he is someone who creates good communities for others. But on his journey Revheim has been forced into psychiatric care and undergone years of therapy as a result of the violence and abuse he suffered up until the age of eighteen. When he began working with Blå, he had just moved out of a psychiatric institution where he had been living for four years.

He doesn't remember the first blow. Nor does his mother. Such violence was so normal, it's now lost for them both in a haze of painful memories; it's hard to separate one incident from the other. The worst episodes, however, are clear: one involving firearms, another where his father tore open a wound on his mother's stitched-up hand, the sound of his mother's screams through the bedroom wall in the middle of the night. But it wasn't these memories that would affect Martin the most, because the worst part was perhaps the unpredictability—the fact that he didn't know when the next episode would occur—and the deliberate malice, the wicked things his father did even when he was totally sober.

"He hit me perhaps fifty times altogether, it's hard to say. And when I was eighteen, it stopped. Had my life started when I turned eighteen, I would have been totally fine today," says Martin, who is now approaching fifty and is married with two children. But he hasn't been totally fine. These first eighteen years have left a mark on him from which he may never be entirely free.

About one in twenty children grow up in circumstances marked by violence on a level classified as pure child abuse. And this ratio has been constant over time. The fact that his father was a respected clergyman no doubt contributed to Martin never being noticed and rescued from the situation. Nobody asked critical questions. No one investigated. Martin desperately hoped that someone would find out what was happening, but nobody rang the doorbell to ask if his father was drinking and hitting. Before packing his schoolbag and leaving the class, Martin would wait until all the other children had left, hoping that the teacher would ask him how he really was.

He hoped the police would ring the doorbell, although knowing that a long investigation would put both him and his mother in great danger, he imagined them locking his father up right away. But he also knew that, in reality, they wouldn't do this, so he never called them. In his teens, he vented his feelings by cutting his arm in the bathroom and watching his blood drip onto the white porcelain, and he developed a generally unhealthy relationship toward food. These are both natural ways of controlling your emotions, when life is otherwise turbulent.

The violence and alcoholism affected Martin's whole family, who lived next to the local prison. Even after Martin was hospitalized following a suicide attempt, people around him didn't realize that what he'd done was more a symptom of an unbearable situation than a specific mental problem. He was actually reacting quite normally to something that was insane, namely that the person he originally turned to for love, security, and validation was also violent and cruel to him. Martin was constantly in danger of being ostracized, considered inadequate, rejected, and being unprotected. But he also loved and admired his father. "I was always ready for Dad," he writes in his book *Å sette sammen bitene* [Putting the pieces together].

Martin remembers incidents where he would get beaten up at school, adding to the humiliation he was experiencing at home. Research shows that children who experience violence and insecurity at home are more likely to be victims of bullying at school. This gives them a double burden: with the frightening situations at home, they have less reason to trust their peers and therefore struggle to bond with them—another effect clearly shown by the Minnesota study. According to the researchers, children who grow up in unsafe conditions will find it more difficult to make friends. The suspicion, uncertainty, and restlessness become part of the child's character and only reinforce their loneliness. It creates a vicious spiral.

Martin Eia-Revheim was an adult when he first began talking about what had happened to him. He still worries that he might lose everything as a result of the most important people in his life suddenly turning against him. There's a potential crisis around every corner.

"If I buy stuff online and it arrives in the wrong size or a screw is missing, it will feel like my life's ruined; everything falls apart, kind of. If my boss calls me in for a chat, I'll be terrified because I normally think it's because I'm going to be fired. I am always worried," he explains.

"Were you lonely as a child?" I ask.

"Yes, I was lonely," he says, "but it's hard to give a precise answer to that question. It's only now, as an adult, that I can see how lonely I was, and on so many levels. But at the time I had no language for it."

He still needs help occasionally, as an inpatient at the psychiatric ward. And to this day he sleeps at night under an extra-heavy duvet, which presses him against the mattress and holds him in place, but even then he never gets more than a few hours' sleep. The fear has such a grip on him that it drives him out of getting any rest. He is always on the go, and wherever he does go, he walks; it's hard to sit still, hard to relinquish control.

"Even when I was little, there was a gulf between my outward identity and what I was experiencing at home," Martin says. "I was two different people. As a child I was constantly lying—to cover up for my father."

And lying shuts you out from the people around you, especially when it concerns something as important as an attack on your integrity, which violence is.

"My mother sees it more clearly," he says. "She says that she was very lonely, that she lost the best years of her life in this destructive relationship. But although it was unclear to me at the time, I now know how my loneliness manifested itself. It was when the school nurse reached out to me and I didn't take her hand. It was the gap between realities. It was when I played football with the boys and never mentioned what was happening at home. It was a powerful sense of being outside everything.

"But no one has really asked me about this before. And there's so much shame associated with being lonely, so I don't like talking about it, of course."

What Marthe's and Martin's stories describe, all too similarly, is normal for the approximately 240,000 Norwegian children who are growing up with alcoholic parents or parents struggling with psychiatric disorders. Twenty-five percent of children in Norway are living in "difficult upbringing conditions," and 20 percent of Norway's 1.2 million children have experienced some form of psychological and/or physical abuse, on at least one occasion, of which 5 percent are exposed to gross physical violence. This means that in Norway gross violence alone affects approximately fifty thousand children annually. These children don't fall asleep to a lullaby, but to the sound of a family in chaos; they wake up in the night to the sounds of violence and rape; they are children who don't bring packed lunches to school, who turn up inadequately dressed in freezing temperatures, children who do the cleaning and washing at home from a very young age, who cover up their cuts and bruises. Countless others have no visible marks on their bodies, but much bigger marks on their souls: shockingly, many children know they are not loved, that they do not belong anywhere, that they are unwanted and in the way. It is a dark and painful loneliness.

"Forty years ago, neglect wasn't considered that important, but now we know a lot about how much and in what ways it affects a child and see that it has both short-term and long-term consequences," says Carolina Øverlien, a researcher at the Norwegian Center for Violence and Traumatic Stress Studies (NKVTS), who has dedicated her life to documenting the effects of violence and neglect.

Øverlien also works as a professor at Stockholm University, specializing in the same field. A while ago, she was the expert behind a Swedish government study that interviewed elderly people aged seventy to eighty, people who had grown up in foster homes or institutions during the 1930s, '40s, and '50s, who had been exposed to violence, sexual abuse, and/or neglect. For some, it was the first time they had ever spoken about the abuse, violence, and the loneliness and anxiety they had experienced. After a lifetime of silence, they opened up about their dark secrets and allowed researchers to see what kind of wounds they had been living with.

"I learned a lot from that study and was amazed by how much damage had been inflicted, what they had experienced in the orphanages," Øverlien says. "And at the same time, I was surprised by how much strength and resilience these people had. Despite everything, many of them had managed to build well-functioning lives, but the taboo was still present. We often say that the taboos around these things have been lifted nowadays, but I find that there is still a huge amount of shame."

For twenty-five years, Øverlien has researched violence against children, how it affects every part of the child's daily life, how it makes the child unfree and afraid, and how this continues well into adulthood.

"What puzzles me is that we know quite a lot about this, and yet it keeps happening, and at such a huge price to the individuals and society," she says. "The knowledge we have just isn't getting out there. During the pandemic, I was especially worried about the kids who weren't attending school, because school can often be the one place where they feel protected, and where their situation is detected and referred to child welfare services."

The children Øverlien is talking about are twice lonely: at home, where they get clear signals from one or both parents that they're not wanted or loved; and in the outside world, where loyalty to their parents and shame prevent them from seeking help. The secrecy also creates a kind of invisible barrier blocking friends and classmates. Children who are exposed to physical and psychological violence struggle to feel understood, because they are unable to share what they're feeling. In addition, these children experience far more toxic and prolonged stress than other children, with all the unfocused and chaotic behavior it causes. This in turn pushes other people away.

But this loneliness doesn't necessarily end up among the statistics Carolina Øverlien uses. If the loneliest children in society are the ones experiencing neglect, it doesn't necessarily mean the researchers detect those cases.

"Growing up unloved, knowing that you shouldn't have been born, and being regularly exposed to offensive and derogatory words will make a child feel incredibly lonely—but this is far more difficult for our research to measure. It's easier to measure the kicks and punches," Øverlien says.

In 1993, researchers approached 126 men, in their fifties and sixties, who had taken part in a survey as university students forty years earlier. The researchers made a startling discovery. Literally all the men who, in the 1950s, reported having an "okay" or "strained and cold" relationship with their parents were now suffering from serious illness, such as cardiovascular disease, high blood pressure, and stomach ulcers—unlike those who had described their relationship with their parents as "very close," of which only half were affected by such illnesses.

You may think this is just a coincidence, because the study had so few participants—but that's not at all the case. And I can say so thanks to the findings of a huge study, carried out in San Diego by what was then an obesity clinic.

When Dr. Vincent Felitti set up his treatment facility for the morbidly obese in 1985, he didn't know that he would also help break down the division between body and soul that had been accepted by modern medicine since the seventeenth century, when the French philosopher René Descartes published his groundbreaking book *Meditations on First Philosophy*. Before Descartes, doctors and healers had been interested mostly in creating a balance of "humours," the four main fluids of the body, believing that if you had too much of one fluid, it would affect your psyche—if you had too much black bile, for example, you were in danger of become melancholic, which says something about the fact that body and soul were considered one and the same. Until the sixteenth century, when surgeons began performing autopsies, the human body was quite mysterious and believed to be governed by mixed fluids and astrological constellations.

Descartes was revolutionary because he described the body as a solely mechanical object, and so during the seventeenth century, scientists began studying the body as though it was a set of mechanical responses. This is known as a "Cartesian" relationship to the body. For example, the circulatory system and the heart were first described by the scientist William Harvey as a soulless pumping system, rather than as the "king of the organs," which was once believed to be linked directly to the sun. This may sound very theoretical, but Cartesian dualism had a very tangible effect on how the body and soul have been treated. Descartes believed, among other things, that one perceives pain depending on how rational one is, which meant that the pain of an animal, child, or woman was taken less seriously than the pain of a white man, because white men were considered the most rational, of course. It was 2022 before any study showed that less research is done on women's diseases than on men's diseases. Even more horrifying, until quite recently the medical world continued to believe that babies don't feel pain when operated on without anesthesia. In 1987, however, a researcher called Kanwaljeet Anand, now a professor at Stanford University, published a scientific article about how much babies experience pain in one of the world's most important medical journals, *The Lancet*. Babies can in fact die of shock if they are subjected to heart surgery without anesthesia, which had regularly been the case. This shows what little importance has been placed on children's psychological and physical pain. And although Bowlby had already described how much hospitalized children need their parents back in the 1950s, the practice of separating sick children from their parents continued for decades.

There are, however, many *good* things about Cartesian dualism, because in the last few centuries medical science has made a string of huge and quite amazing breakthroughs by seeing the body as a kind of machine that can be fixed. Today, we no longer fear diseases that once threatened us with death and mutilation. We have vaccines that protect us against measles and polio and Covid-19. We can treat cancer and tuberculosis and heart disease. But while getting to this point, we lost something. According to Descartes's model, the psyche has nothing to do with the body, and so treating the body and treating the soul have for centuries been two completely separate disciplines.

Dr. Felitti didn't realize he would be helping to shake up this paradigm when he began treating morbidly obese people from San Diego's middle class, people who of course knew that broccoli and exercise are healthy, and that overeating and drinking soft drinks are unhealthy, and who, with Felitti's guidance, managed to lose weight quite easily. However, as soon as their weight-loss program finished, all the weight came back. Eventually, Felitti grew frustrated with treating people who repeatedly became obese, so he decided to take a closer look. He began by interviewing over two hundred of his former patients and made a startling discovery. Fifty-five percent of the obese people he had treated had been sexually assaulted on at least one occasion. It's not surprising that he noticed this, considering that on average "only" about 10 percent of women experience sexual abuse. Was there a connection? It also became clear to Felitti that his patients had not gained weight slowly; they had become obese suddenly, as a direct consequence of a traumatic event.

"Nutrition has nothing to do with obesity. Teaching people about nutrition is something we do because we assume people put on weight because they don't know any better," he said after examining thousands of overweight people.

Because people do know better. And whatever these people were suffering from, it certainly wasn't ignorance. Dr. Felitti's research eventually produced a list of childhood traumas, which he called ACES (adverse childhood experiences). If you can tick six or more of the traumas on Felitti's list, your chances of developing some kind of addiction—such as an addiction to food, where you use food to regulate emotions in adulthood—increase by 4,600 *percent*. Eventually, Felitti found numerous other physical consequences of childhood trauma, such as heightened risks of heart attack and cancer. And as he continued to investigate, he discovered a large and significant connection between your experience in childhood and your physical health afterward. Yes, the telomeres, which are in our genes and affect how long we live, get shortened by the extreme stress of complex PTSD, for example. Basically, when trauma eats into your development from such an early age, you end up with a shorter life.

At roughly the same time that Vincent Felitti created his ACE study, researcher and doctor Anna Luise Kirkengen made the same discovery.

"Our immune system is affected by our experience of violation and abuse," she says in an interview. "If the victim is left alone with their experience, the result is colossally destructive. By the same token, I'm convinced that if the person in question gets the right help, some of this disease's development can be prevented."

Toxic stress harms both the body and the brain. Many experts want developmental trauma disorder and complex PTSD to be included in the *Diagnostic and Statistical Manual of Mental Disorders*. "Chronic trauma changes the entire neurobiological development and the capacity to integrate sensory, emotional and cognitive information into one comprehensive experience," writes researcher Bessel van der Kolk, who has worked exclusively with trauma. "Developmental trauma lays the foundation for unfocused responses to persistent stress, and burdens the healthcare system, prisons and any service related to social support and mental health."

Talking about chronic stress sheds new light on the relationship between body and soul. Long-term stress creates low-intensity inflammation, which in turn breaks down the body and causes physical and mental illness.

Children who have lived in a constantly threatening situation will operate in the autonomic nervous system called the sympathicus. Researchers have found that even babies react noticeably to stress, created by the sound of an argument in the room next door, for example. Since the baby has no other defense mechanism at hand, its only survival response will be to completely freeze. Traumatized infants will become used to freezing quite often. But over time, being in the sympathicus isn't particularly good, and over years it triggers low-intensity inflammation, which in turn leads to serious autoimmune diseases such as rheumatism, diabetes or fibromyalgia; insomnia, poor digestion, concentration issues, anxiety and depression; and a high risk of cardiovascular diseases and cancer. In recent years, research on inflammation has taken off, and the findings show that low-intensity inflammation from toxic stress is extremely harmful to us.

Researchers also believe that extreme stress does something to the brain and nervous system. It weakens the connections in the frontal lobe, where the executive functions—like paying attention and concentrating, solving external tasks, estimating consequences, and having a clear plan and direction—are processed. The hippocampus (memory center) and the amygdala (emotional center) both function less well in children who have experienced repeated trauma and neglect while growing up. And this trauma doesn't even have to involve physical violence and abuse. Brain scans revealed that children who felt neglected and unsupported, who were bullied and ridiculed by their parents, or who regularly experienced not being cared for and understood had weaker connections too. And these weaker connections can lead to poorer concentration and memory. This means that the brain cannot distinguish between real danger and imagined danger, which puts the body on constant alert. The brain then becomes less able to regulate stress and handle negative experiences. In the Minnesota study, the researchers saw how children with insecure, and especially disorganized, attachment patterns are far more affected by negative stress and take longer to recover after an upsetting incident than children with secure attachment patterns.

The things that suppress inflammation—and are essential to good health in general—are nevertheless basic: community, friendship, and a sense of belonging and meaning are strongly anti-inflammatory. And the encounters we have with the greatness of nature—with sunsets and sunrises, rainbows, and stars scattered across the dark night sky—have a calming effect, and make us feel small, in a good way. It's a feeling Moominpappa describes in *Moominpappa at Sea*: "Sometimes, when night drew closer over the sea, I liked to take over from Hodgkins at the helm. The moonlit deck that slowly rose and fell before me, the silence and the restless waves and clouds, and the solemn circle of the horizon—everything gave me the nice, exciting feeling of being terribly important and terribly small at the same time (perhaps, however, more the former)."

When we are by the sea, research shows, we are in the place that makes us happiest. One large study, called BluHealth, has investigated the effects the sea has on the psyche; standing by the sea activates the brain system that researchers call the default mode network (DMN)—daydream mode—where introspection occurs, and where we process our memories and relationships with other people. The research shows that when we daydream, we have good ideas, and it is also the time when we find out who we are in the world and clearly envisage the future. Visions of the future give us hope.

Researchers have found that when we operate in sympathicus, the brain bypasses its higher functions and goes straight to our animal responses, to the limbic system. By doing so, it limits our creativity and perceived happiness, our sense of being free and happy. An abused child isn't free to act based on its needs and wishes, nor can it play or daydream without being constantly afraid. Neuroscientists call this the "hijacked self." The daydreams and the free leaps of thought we have in DMN mode are crucial to the development of the self and our relationships with other people, yes, for finding direction and meaning in our lives. The DMN is often linked to positive time spent alone, when you wander around lost in thought, just like Snufkin, jumping freely from association to association. But if you have been traumatized as a child, this neurological system gets altered; the DMN doesn't get activated when you are relaxed and calm, but instead when you are totally alone, worried that you are in danger, afraid that no one is looking after you.

It gives you a totally altered sense of self. This network— which for others is a private space, somewhere for testing ideas and processing feelings—is for traumatized children directly linked to depression and fear. Researchers have long wondered where rumination—the downward spiral of brooding and negative thought—comes from, because it's in one way or another connected to the brain's DMN system. Rumination is related to depression, which is perhaps the connection. Studies show that lonely people are more often in DMN mode than are people who are not lonely. But for lonely people, there's nothing positive about it.

Children who never had a safe upbringing will also struggle with emotional self-regulation; regulating our emotions is something we learn from secure relationships. People who don't learn this develop problems with so-called mentalization: they find emotions hard to understand and deal with; they become captive to their emotions and unable to calm themselves down and struggle to understand others—which can in turn make them irritable and aggressive. All this makes connecting with your surroundings, listening to others, and being empathetic and present more challenging. You become harassed, unfocused, frightened, and stressed out, someone who finds it harder to make friends. Learning, self-reflection, and the so-called executive function are also affected by PTSD: the executive function concerns our attentiveness and consequence thinking and is important for what we call working memory. If it gets damaged or disturbed, we become less focused and attentive. That is why it is so easy to believe that the symptoms of children who have been abused and exposed to violence are due to ADHD.

Researchers working with traumatized children describe a "window of tolerance" where we are focused and present, somewhere between being under- and overactivated. But it's a somewhere that children who experience severe violence and neglect rarely find themselves. They are either restless and hyperactive, ready to run, frightened and fidgety, or they go into what's called "hypoactivation," which is when they are in fright mode and completely switched off.

The child who curls up and disappears is feeling so lonely that it seems pointless to even signal for help. Total surrender and vanishing into the mind have become the child's only possible escape route.

But if you don't know better, you don't properly understand. The child's fear responses can easily be classified as ADHD or depression, or the large swings between fight-or-flight and fright can be interpreted as bipolar disorder. If those treating the child don't know what they are seeing, they misinterpret the signals. One study shows that ADHD is the most frequently given diagnosis to sexually abused children. Neglect is often misdiagnosed as ADHD as well. So, now that we know that stressed and traumatized children with PTSD will zone out or feel extremely restless and afraid, it's not surprising that these children are unable to concentrate well enough to keep up at school.

For several years, Inga Marte Thorkildsen was the minister for schools in Oslo municipality, and she is highly concerned about these very children. During the Covid-19 pandemic, the number of alerts received by child services fell dramatically. All the places where a child's well-being might get noticed were closed, and there were periods when children were unable to attend school or visit the school nurse. The organization Mental Helse [Mental Health Norway] is now reporting a record number of inquiries from children and young adults, and in year two of the pandemic, first-time inquiries from children and young adults increased as well. Vulnerable families struggled even more during the quarantine periods. Those helping people with eating disorders have reported an explosion in demand for treatment.

But one problem existed long before the pandemic: our fear of tackling a sensitive issue when everyone around a child is too afraid to ask. Often, those who find themselves in close contact with an abused child are quite unsure *what* to ask, or they do not know how to resolve the situation if what's really going on in the child's home becomes clear to them.

"It's easier to talk about eating disorders or finances than it is to talk about sexual abuse and violence. The symptoms are easier to deal with than the causes," Thorkildsen says. "Teachers learn far too little about this when they're training, so when children tell them stories, they don't know what to do. We also have this notion of treating these children 'professionally,' which often means sending them to a person they have no relationship with, who works according to strict treatment guidelines and perhaps uses methods the children themselves find prescriptive and alienating. It doesn't give children the confidence they need to really share their experiences with the adults around them."

Thorkildsen is convinced this is the wrong way of doing things. The way out of loneliness is about trust and love.

"The most important tool when dealing with a child who experiences violence and abuse must be to show them warmth and love, to be open and accessible and flexible based on the individual child's needs," she says. "Unless they are cared for in this way, children will close themselves off and not say anything."

During the pandemic, Thorkildsen personally helped and supported teenagers who were experiencing violence and sexual abuse. "The big problem when dealing with children who are carrying such experiences is that adults don't give children sufficient time to find the courage to fully unpack everything," she explains. "They'll listen but will often prematurely think they've understood the whole picture. They may then ask other adults to double-check the information, without even telling the child. This not only puts the child at risk of being punished, but it also destroys the child's faith in adults being their protectors. The abusers' defense mechanisms are enabled by a system that cooperates too uncritically with parents, because the systems themselves are so under-resourced and the professional training is extremely inadequate."

Knowing that their health and safety isn't worth looking after will destroy children's core trust in their surroundings. And it's a feeling that stays with them as adults. It ruins their lives. It undermines their faith in institutions, their faith in everyone with power, it eats away at democracy from within.

So, what is on the list of childhood traumas that Vincent Felitti's research uncovered? Felitti found that adverse childhood experiences (ACEs) result from growing up with adults who take drugs or have mental illness, from experiencing the suicide of a close family member, from witnessing or experiencing violence, from being exposed to abuse or neglect. Unstable homes or absent parents—sudden death or imprisonment—also play a part in creating insecure conditions for a child's upbringing. In a survey of twenty-five US states, 61 percent of adults reported that they had been exposed to one or more of the experiences listed by the ACE study. Divorce is also on the ACE list, but before you get worked up about Felitti's conservatism, remember: before a separation occurs, a child is often surrounded by a great deal of turmoil and conflict and will then experience a totally different living situation, which will also affect health and stress levels later on. All this puts a burden on a child, resulting in anxiety, fear, depression, inability to sleep properly, and spirals of toxic stress.

According to the US Centers for Disease Control and Prevention, approximately 1.9 million cases of heart disease and 21 million cases of depression could have been avoided in the US if the people concerned had grown up in safe circumstances. A rough estimate puts the cost of all this violence and abuse on American society, from illness and lost manpower, at $500 billion. I haven't found a corresponding figure for Norway, but we can assume it's significant. Nevertheless, both lives and money could have been saved had children like Martin Eia-Revheim and Marthe Bødtker been unburdened of their secrets, had they not been in the absurd situation of having to cover up for their adult abusers while constantly fearing violence or rejection from their closest family.

Even those who tick just one of the boxes on the ACE list have a much higher risk of suffering from chronic depression—10 percent of men and 18 percent of women. For anyone who can tick four or more boxes, this likelihood increases to 33 percent of men and 60 percent of women. In general, women react even more strongly to childhood trauma than men. Women suffer more often from anxiety and depression, and they more often develop autoimmune disorders, such as rheumatism and lupus.

Since it was published in 1998, the first ACE study has been cited seventeen thousand times and has transformed the field of trauma research. Now the consequences of child abuse, both psychological and physical, are taught and researched all over the world. It was impossible to ignore what Vincent Felitti had discovered: the first ACE study involved almost ten thousand people and proved once and for all that the psyche and the body are by no means compartmentalized. The extreme and toxic stress we experience leaves clear physical traces.

"The strange thing was that we didn't need to do much to bring our obese patients' weight down," Felitti says today. "Just listening to their stories was enough. It had such an impact."

It was like a modern confessional: doing something as simple as listening and making a patient feel understood could change their body and psyche dramatically.

What's strange is that the ACE study shows that a traumatic childhood leaves the same physical traces as those identified by loneliness researcher John Cacioppo: the results of chronic loneliness are identical to the results of adverse childhood experiences. What happens if you are chronically lonely? You become depressed and anxious, you overeat, you resort to drugs and alcohol, you mistrust your surroundings and develop attachment issues, you appear less empathic, you become hypervigilant, you develop sleep problems, heart and vascular disorders, diabetes, autoimmune diseases. I feel like I'm reading the same statistics, like I'm comparing two finished puzzles and realizing that they're pictures of the same haunted castle, but from different angles. And while I can't immediately assume that these two sets of statistics coincide and describe the same people, it seems very likely. One of the questions in the ACE test is "Did you often or very often feel that ... a) No one in your family loved you or thought you were important or special? or b) Your family didn't look out for each other, feel close to each other, or support each other?" There is no other way to look at it: this question is about loneliness, about whether you feel safe in your most important flock. And this one point, about being loved and considered an important part of a safe family, can be so important that it will overshadow everything else. One large study showed that the feeling of being *unloved* played a bigger part in the development of eating disorders than physical violence or rape.

If, as a child, you feel unloved and uncared for, if you don't feel even a basic level of security or trust the people and the world around you—won't that lead you into a lifelong sense of loneliness? When loneliness researchers describe the huge numbers of lonely people in the world as being almost a mystery, a sudden phenomenon, a virus, a pandemic, is it perhaps because they can't see where loneliness comes from? Yes, many incidences of loneliness are transient and situational, according to the research; we can get lonely in all the transitional phases of life, as when changing schools or starting a new job, moving to a new house or into an institution. But for many people, loneliness is chronic and persistent. And I believe that loneliness comes from somewhere. I think lonely children become lonely adults. I think lonely adults become lonely seniors; though I don't just *think* it, I'm absolutely sure of it. I was one such child.

"I began looking at elderly care," says Inga Marte Thorkildsen, "and it struck me that the violence, trauma, and difficult relationships these patients experienced in life play a role in the development of dementia. Toxic stress can lead to dementia, and when these people 'revert back their childhood,' they also remember the traumatic events of their childhood, which they have spent a lifetime trying to escape. Note that an enormous amount of force is used in geriatric psychiatry."

A high ACE score means you are more likely to develop dementia. The loneliness an elderly person has experienced also increases their risk of developing dementia, though I can't know whether the elderly person had felt this loneliness since childhood. (Again, it's as if I'm looking at the same picture from two different angles.) A research group studied over eight hundred elderly people, following them over a period of four years. And the risk of developing Alzheimer's doubled for those who felt lonely.

Inga Marte Thorkildsen has, as mentioned, worked to change the situation for children who have been exposed to violence in Oslo. She believes that these children need to be treated with far more humanity, more empathy and understanding.

I know a lot about what I'm not going to pass on to my child: fear of the outside world—she'll never learn that from me. I never want her to feel that she can't approach her own parents with her problems. My gift to her will be trust. Yes, I've been the Groke, but it doesn't have to be like that. My child will not learn to withdraw from other people.

"The social strategy that loneliness induces—high in social avoidance, low in social approach—also predicts future loneliness," writes John Cacioppo in his seminal book *Loneliness*. It becomes a vicious spiral. I have been in that spiral.

When you don't trust other people, even friendly compliments don't sound friendly but instead adopt a sheen of irony. It's hard to believe that others mean you well, that someone wants to offer you care and love. When you are unable to accept true love when you encounter it, you become more and more isolated.

"When these negative feelings drive the self, when this becomes the very core of the self, it draws you towards states that remind you of the bad ones, destructive situations and relationships," says Mari Bræin, an expert on trauma.

Bræin specializes in PTSD and complex trauma in children. She has also worked with UNICEF on measures for marginalized children and as a special adviser at the Regional Resource Center on Violence, Traumatic Stress, and Suicide Prevention in Eastern Norway. She currently works at the Center for Stress and Trauma Psychology in Oslo.

Many children who have experienced neglect struggle with a so-called intimacy barrier. They can handle being in a new foster home quite well initially—when the relationship isn't that close. But as the foster parents get closer to them, the relationship becomes scary; doubt and fear come into play, there is more to lose, and the child has no confidence in being taken care of in the long run. When you know that you are going to be rejected—just not *when*—you feel inclined to hasten the rejection, to give yourself some control over it.

"We can say—as the German philosopher Arthur Schopenhauer did—that this is the 'hedgehog's dilemma': the closer you get to the hedgehog, the more it raises its spikes," Bræin says. "Schopenhauer opted out of intimacy himself and, as I understand it, was strongly rejected by his mother. He chose music and art instead of human contact."

In other words, many things contribute to making an insecure child behave in a lonely manner and withdraw from other people. Attachment psychologist Stig Torsteinson points out the leap of faith required for us to make friends, grow, and develop.

"Development requires taking chances and challenging yourself," he says. "And if you know there's a safe base you can return to, taking such chances isn't that scary, but if there's no one to return to for support, then an insecure child can become very reluctant to take developmental leaps during relationships."

Torsteinson works alongside psychologist Ida Brandtzæg and has met a lot of children with attachment problems.

"What's clear about children with insecure attachment experiences is that they have imprecise ways of relating to adults," he observes. "A child with safe experiences will go in a straight line to an adult if he needs help. It has taught itself how to ensure that its needs are looked after. Such children expect care. A very insecure child may have experiences that make it afraid of its own parents, perhaps one or both, which makes turning to them difficult when it needs help and comfort. And that's probably the most acute form of loneliness—being afraid of your own parents."

"Children who experience this kind of loneliness will reinforce their loneliness because the experiences of their own parents will have created expectations about how they should be with other people. So these children are less likely to ask for help or comfort when they need it," say the two psychologists.

It's important to remember that children should develop not only the ability to manage everything in their lives themselves but also the ability to ask for other people's support. We humans need good strategies for being in relationships with other people, when required.

"Children with an insecure attachment style are more vulnerable under stress, because they haven't established good strategies for getting help and support," Brandtzæg and Torsteinson point out. "This results in them trying to handle stress and emotions on their own, which is too big a task for a small child, and ultimately gives them prolonged stress and anxiety."

One such neglected child who slipped through the net was social geographer and author Anne Bitsch.

"At the age of thirteen, I went to child services for help, and wasn't taken seriously," Bitsch says in her autobiography *Går du nå, er du ikke lenger min datter* [Go now and you'll no longer be my daughter]. "I managed to speak to a case manager, who then informed my mother that I'd been there. My mother became furious and extremely violent. And I lost all faith in adults and in the system."

Bitsch writes about a bipolar and alcoholic mother and about a father who abuses her when she is thirteen years old. At the end of the book, she confronts all the neighbors and teachers who had known what was going on, asking them pointedly why they didn't do more for her when she was a child. Their answers are evasive, self-righteous, halfhearted. They talk about "untimely intervention." As an adult, Bitsch is now a prominent social commentator and researcher on rape and violence.

"During my research I've seen that in many cases of domestic violence children are not sufficiently cared for, even when they show clear signs of neglect," she says. "Very few cases from the upper levels of society ever reach child services or the justice sector. When it comes to child neglect, there's a big hole in the public debate—a lack of class perspective."

Our notions of violence against children are full of clichés and stereotypes. We don't see the children who are exposed to violence, both because they are so good at hiding it and because it's unthinkable that a woman can be violent and neglectful, for example, or that well-educated people with beautiful homes are capable of behaving so uncivilized with their own children. Most of us are blind to the violence.

"As a child, I genuinely never understood why other people wouldn't help us. And when I think about it today, I still don't understand," Bitsch writes in her book.

She has written a book about the dark, pulsating loneliness that comes from bearing heavy secrets and too much responsibility. It is about the loneliness that comes from the violence and sexual abuse hidden by a mother's facade of a seemingly well-functioning and well-formulated adult life, and from the failure of other adults to act.

"Throughout my childhood, nobody ever asked me: How are you *really?*" she says. "Many people are afraid of interfering and have fixed ideas about who the decent people are and who the abusers and drug addicts are. Society's institutions reflect and bolster class and gender differences. I think there's a deeply ingrained reluctance for those in the protection services and legal system to intervene against people who look like themselves. Society needs to learn how to look beyond the facade and see how people are really feeling."

Some things are statistically true about violence against children: it is more serious in low-income families. However, it doesn't mean that acts of physical and psychological violence don't occur in all kinds of homes or aren't committed by both mothers and fathers. In fact, the gender distribution is fifty-fifty— many don't see the women who perpetrate violence, because it conflicts with their gender-role expectations. In higher levels of society, the violence is possibly just more hidden, and the violence perpetrated by women is more likely to be psychological than physical. But the violence is there. The violence leaves its mark. Children who are exposed to violence become adults with attachment issues, adults who struggle to form close bonds with the people around them, who don't trust society as a whole. They become Toffle, who was so afraid. They become the invisible child. They become the Groke.

Calling myself the Groke might seem a bit odd. Most of the people who know me find me warm and approachable. Few of them experience what I'm like in a very close, long-term relationship: the longer I know someone, and the more intimate the relationship becomes, the more certain I become that it will end with me being abandoned. I create pointless arguments to prove that's the case, or tests that are impossible to pass. It all becomes evidence that I'm not really loved. Rejection and pain are always lurking, and I want to get there first. I'm very often depressed, always convinced that the world doesn't want me. And it's a feeling I continually have: a fear of not belonging anywhere. I take Vincent Felitti's ACE test: I get five ACES. I take it one more time. Still five. Does that mean I, too, am condemned to a life of autoimmune diseases, insomnia, and depression?

My work on this book has made it clearer to me how I create my own loneliness, simply because I want control, because I don't have any trust. I rarely feel safe. I sleep badly and restlessly—in the past, I always woke up and lay awake for hours in the middle of the night; now it happens less often. Perhaps that means I've gotten better. But during the Covid-19 pandemic something was amplified, the sense of being trapped in my own home perhaps, a feeling I know all too well. As a rule, when this feeling comes, I go for a walk to get some fresh air. Meeting a friend usually solves the problem, or I've trained the anxiety from my body at the yoga studio, or I've just sat alone in a café when the darkness descends, this uneasiness that should be called by its proper name: loneliness. But during the pandemic, I had no means of escape. I had to stay at home. And I resorted to an escape route I haven't used in years: food. As a teenager I suffered from anorexia, but as an adult I struggle with overeating, which, according to the health authorities, is the world's most common eating disorder. Binge eating disorder is driven by the same thought pattern as anorexia. I will think: "I'm so disgustingly fat, I must stop eating, I need some control. I should eat no more than a thin slice of beetroot a day, like I did when I was fifteen (and on the verge of anorexia)." But I'm unable to keep up this regime for more than one day at a time, before I lapse into an orgy of overeating. Food comforts me, calms my anxiety, makes everything manageable. I swallow my loneliness with red wine and macaroni. So, during the two years of the pandemic, I gained weight, relapse after relapse, and each time descended into more self-loathing. "Why can't I stop eating?" I thought furiously. "What's wrong with me?" I know that loneliness increases my appetite for sugar and fat, but that knowledge doesn't stop me from shoveling it down, hungry for comfort.

The sense that I'm not worth loving is so entrenched that it's hard to grapple with; it just sits like a dull murmur. I buy new dresses, far too many clothes; I want to just cover myself and look like someone that other people might like. I work ceaselessly, because I don't trust that other people or social services will care for or support me financially. I won't even trust that my husband has bought the toilet rolls he promised to buy, so I'll buy them myself and end up with duplicate packs of toilet roll cluttering up the bathroom. I'm then ashamed, because I know that I'm behaving irrationally—a type of irrationality that compels you to spend and makes you write glowingly about yourself on social media. Look at me, acknowledge me!

I read books and call books my friends, which is a relationship that's easier to control since it only has one active participant; I've never been rejected by a book. When I was a child, books were my only friends, and I had many, many such friends. Now I write books to get close to people, although readers are people I hardly ever meet. Writing a book is a long detour to love, just as the food I cook and the gifts I buy are long detours to a hug. As a child, I also read any superhero comic as though it applied to me, which it of course did. Like the superheroes, I had a dark secret. Like them, I had a superpower: I could write.

And yes, I'm of course warm and approachable when you first meet me: I'm an expert at connecting with people, lots of people, preferably one auditorium at a time. It's a skill I started learning at the age of thirteen when I started "living" with my only two friends; I adopted their families. My survival strategy is friendship, but without ever relying too much on one person—there's always a risk of me being dropped at some point and needing several other friends to fall back on. I am the perfect guest, but a bad permanent resident. I am always looking for reserves and spare parts. I behave as though I'm totally self-reliant and about to set off on a yearlong solo expedition to the South Pole: I stockpile food and buy too many clothes; I have large reserves so that I can always survive entirely on my own. I never present myself as weak. I never ask for help. I don't trust that people like me for who I am, that I'm lovable, that I belong.

So how did the Groke stop being so lonely? Each night, in *Moominpappa at Sea*, this ice-cold lady comes to stare at the Moomin family's hurricane lamp, which Moomintroll has placed on the beach to keep her away from the house. She looks at the lamp, sings a mysterious song, then turns and goes home. Perhaps looking into the lamp is her way of getting closer to the Moomin family, a detour to intimacy, a typical behavior of those marked by attachment injuries. We, of course, know that the Groke has been neglected since she was little! One evening, however, the Moomins run out of paraffin, but Moomintroll goes down to the beach anyway, albeit without the lantern. And the Groke is so pleased to see him, so intensely happy that he has come down to the beach purely to see her. Their ritual with the storm lamp has given her some predictability and security. Moomintroll has created something magical: he has established trust. And they have now become friends.

"Moomintroll moved forward in amazement. There was no doubt about it: the Groke was pleased to see him. She didn't mind about the hurricane lamp. She was delighted that he had come to meet her. He stood quite still until she had finished her dance. He saw her shuffle off down the beach and disappear. He went and felt the sand where she had stood. It wasn't frozen hard at all, but felt the same as it always did."

I'm sure that I'm warmer now, that I've stared long enough at the hurricane lamp that I can dare to trust other people. And I wonder if I should become another person, but no, I should live with my wounds; after all, the Groke wasn't transformed entirely, she is still the Groke. I also decide that I'll never teach my child my survival strategies, because they are lonely strategies. And since I know that my daughter mirrors me, right down to my tiniest movements and most advanced swear words, I realize that I have to adjust my faith in the world. I'm going to stop taking detours to love, I'm going to go straight toward it. Sometimes it seems near impossible. But sometimes—when my smiling seven-year-old curls up beside me, ready for her bedtime story, for example—it's no problem at all.

As I read about the Groke to my daughter, lying safely in the crook of my arm, something about the story becomes clear to me. Although the Groke is dancing and radiating warmth for the first time in her life, there's one thing she doesn't do: she doesn't offer Moomintroll a hug.

7

Why we can yearn so badly for a hug

WE ARE ALL created by a hug. Not the hug that brings egg cells and sperm cells together; I'm thinking of the one you grow into. You didn't know it at the time, but you were held tightly from the first second of your life. You were firmly attached to, and gently squeezed by, a pulsating body. And you probably didn't notice how cramped things were until your leg started pressing hard against the side of that little cavity, or when you felt surrounded and had no room to move. Like a weird kind of bear hug, or a straitjacket. Before long, you're squeezed even harder, then spat out into a totally new world, much bigger than the one you came from. And the first thing you encounter is probably another hug. Someone holds you tight and gives you warmth, skin contact, and food. You wouldn't survive otherwise.

Skin is the largest and one of the most important organs in the body. And the very first sense we develop is touch. A fetus doesn't develop taste and smell until its fourth month. But two months after conception, a fetus will react when it embraces itself or touches its mouth. It is there, in the womb, that we first feel something. An unborn child touches its mouth constantly, throughout the rest of the pregnancy, and quickly becomes attuned to the sense of touch all over its body, like a plant waiting for water.

To illustrate how vitally important skin contact is, I must tell a horrible story from 1958. At the University of Wisconsin–Madison, psychologist Harry Harlow had made a strange observation: baby monkeys, when separated from their mothers right after birth, would become especially attached to the blanket in their cage. To investigate this more closely, Harlow set up an experiment. He placed the almost newborn monkeys in a cage with two "mothers"—one made of metal but with a feeding bottle full of milk, the other made of soft teddy-bear fabric but with no bottle of milk. The monkey babies living in these cages spent as little time as possible with the metal mother, and instead spent most of their time clinging to the teddy mother. They would run only briefly to the metal mother for a quick drink of milk, before running back to the safe arms of the teddy mother. What the experiment showed was something that ought to be obvious to most of us: cuddling and closeness are at least as important for babies as nutrition, perhaps more important. Harlow subsequently spent his entire research career pursuing this lead, torturing baby monkeys in every possible way to find out something we instinctively know: We are social animals. We need closeness.

Another equally horrible "experiment" made it forever clear what growing up without cuddles and physical closeness does to humans. When the former shoemaker's apprentice Nicolae

Ceaușescu came to power in Romania, a central part of his pol-
icy was the banning of contraception. Women should have as
many children as possible. If they had ten or more, they were
even specially thanked and commended. The dictator believed
that having so many children, so many workers, would be a
boost for the economy. But what he didn't have any plan for
were children who were sick or malformed, with intellectual
disabilities, or with something as common as poor eyesight.
Such children were considered defective and handed over to the
authorities, who locked them up in large, understaffed institutions where they received the minimum amount of care. Most
were confined to their beds behind barred windows. These children never played, were never given a good-night hug or kiss on
the forehead, never held or comforted when they fell over. They
never learned to speak, never learned the word which, for most
people in the world, is the first: *mommy*.

When the dictatorship collapsed in 1990, Romania's institutions were opened up to the world, and altogether 170,000
children were found living in appalling conditions. The first
outsiders to enter these houses of horror saw injured and mal-
nourished children lying in pools of their own urine; they were
underdeveloped, self-harming, or rocking back and forth, crying
and screaming hysterically.

In 1998, researchers compared the film clips of Ceaușescu's
orphanages with footage of Harlow's baby monkeys. The similarities were striking. But it shouldn't have surprised them. John
Bowlby had already studied hospitalized children, who were
found to be miserable and to take a very long time to get well,
despite being given all the required nourishment. This may
seem a little hard to understand: Wasn't getting food and clean
sheets enough for a child in the postwar period? At the time,

however, Bowlby launched a groundbreaking theory: children need an attachment figure, a parent or carer, who can give them something entirely different from just food.

But if we go even further back, we can see that they knew this already. During the Second World War, psychologist Anna Freud found herself with a natural experiment on her hands when many of the children living in London were moved, for safety, to the institution she had built in the suburb of Hampstead. Surprisingly, the children who were allowed to remain with their parents in London fared better; it turned out that being deprived of their parents was far worse for the children than experiencing a rain of bombs. It didn't matter that it was done for their own good, because the separation had a distinct effect on them. Yes, I know, it seems almost comical that researchers needed to say something so obvious: children need cuddles and closeness from caregivers. So it turns out that a person can offer a child something irreplaceable. At the same time, babies in the 1940s were seen to have a high mortality rate, even in orphanages and hospitals that were tidy and hygienic. Harry Bakwin, professor of pediatrics at Bellevue Hospital in New York, believed that depriving children of hugs made them lonely, so he encouraged the nurses to cuddle and play with young patients in the children's ward. The death rate at Bellevue was low compared with that at hospitals where the babies and toddlers were not cuddled for fear of infecting them. Not being cuddled is fatal.

Body contact is an essential part of how we become human, and the most intimate way of having contact with other people is via the skin. Touch can register pain and comfort, warmth and cold, and is vitally important to finding our place in society and the world in general. Our skin is the border between us and the world, our point of contact with other people. This means that there is an entirely wordless form of loneliness, one that is difficult to spot: skin hunger.

"We need to be touched," says trauma expert Mari Bræin. "That initial and most fundamental form of trust helps shape us. As a baby, you've been inside this warm and safe pocket, and then you emerge into the world and register your transition into life and light. You notice whether it's safe, whether the breathing and heartbeat of your mother or the father who takes you is calm or flustered. Is it nice being held, are the hands calm and warm, or are the movements jerky and abrupt? How you are held and taken care of at the very start at the start of your life is crucial. It's where it all begins."

In Bræin's work with traumatized children, she initiates contact via touch when words aren't enough or aren't there, using a method inspired by the researcher Bruce Perry, among others. Exploring loneliness means working physically, because childhood experiences of neglect remain in the body, making the loneliness a physical and totally wordless experience. It's a feeling of intense, physical vulnerability and helplessness, a sense of not being protected and held. Touch is absolutely essential to our trust and connection to the people around us and is thus directly connected to loneliness.

"Most people have a form of existential loneliness," Bræin says. "It's a crisis you arrive at in life, when you realize that it doesn't matter if you live or die, that you're just a very small and meaningless part of this big world. But then there are children whose lives actually begin like that, who carry this realization from the very start: existential loneliness is their permanent setting, and it colors everything in their world. What for others is a passing crisis can for them be their permanent view on life."

When you experience this kind of loneliness, you are sure that you cannot be part of a community, that you don't deserve help or solace, that you've been abandoned, that you don't belong anywhere. It is a disposition that can be caused by violence and abuse. But the "mere" absence of love and touch can also lay the foundation for a lifelong feeling of existential loneliness.

Brain development in our initial months and years is directly linked to the warm contact we have with our caregivers and the world around us. Research on the kangaroo method clearly shows this: premature babies who spend their first months in constant skin-to-skin contact fare relatively well when compared with similarly premature babies who have been placed in an incubator. Babies cared for using the kangaroo method lie skin-to-skin with their parents for the first months of life, feeling their parents' breathing and heartbeat, and greatly prefer this warmth and coziness to being in a sterile box. An American study from 1986 also showed clearly that premature babies who were regularly massaged grew faster and were ready to be discharged from hospital much earlier than those who were not massaged. Another study showed that the sooner a newborn baby makes skin-to-skin contact with the mother, the more likely it is that the mother will be successful in her first attempt at breastfeeding. And while blood sugar levels and heart and lung functions were better in the children who did receive immediate skin contact, the stress associated with a lack of skin-to-skin contact and cuddling was something all children could experience.

"A study of children in nursery institutions found that 'children in daycare exhibit higher cortisol levels than children at home,'" says Francis McGlone, a professor at Liverpool John Moores University and one of the world's leading experts on touch. "The study concluded 40 percent were stressed. In the

US they do something very worrying for me: they place four-month-old babies in kindergartens; sometimes they are even younger. It is stressful for them not to be held and cuddled, usually by their mother, such as during breastfeeding, at this critical stage at the beginning of their life."

The stress hormone cortisol is a driver of toxic stress and low-intensity inflammation—both of which I've discussed above, and both of which are fueled by emotional stress. Being stroked and fed by one or two clear caregivers as an infant can be key to our ability to regulate emotions—that is, our ability to understand and manage emotions.

If I don't get enough of this, it affects my whole body; I'll long so badly for a hug that it hurts. My friendships involve long hugs, and the pandemic deprived me of them. Meeting someone on a screen cannot satisfy that need; talking isn't enough. To go without physical contact is anxiety-inducing and painful.

Social anthropologist and biology professor Robin Dunbar believes we can learn more about friendship from other primates. Apes spend 10 percent of their waking hours grooming each other. And not for the sake of beauty or because they are bothered by insects—Dunbar has found that when apes pick through each other's fur, it is mostly a social activity. They are stroking and scratching each other's backs—and it's here that friendship begins. This mutual cozying is one of the keys to understanding our sociality and to understanding the life-threatening stress that hits the body the moment we get lonely, when our bodies are no longer touched.

So, when the pandemic forced us to remain one meter apart, what did we do? Even well-adjusted people in secure relationships felt the strain of the isolation it caused. Perhaps that's why loneliness started eating into the population—even people with friends, loving parents, people who never experienced violence, discrimination, bullying, or racism, even they felt the pandemic as a physical pressure as we entered its second year. Unfortunately, self-touching cannot totally replace the touch of another person—the quite surprising and unpredictable feeling of someone else's warm hands and body is hard to give yourself. But it does help to give yourself a good hug every now and then. And it's possible that being close to something warm, or taking a hot shower, can give you a sense of security and of being less lonely; it can create the illusion of lying beside a warm caregiver's body.

Some studies show that lonely people seek warmth more often. There are many ways to be touched, and in some countries you can even buy hugs. In Portland, Oregon, Samantha Hess once built a thriving business as a professional hugger. On her website Cuddleuptome.com, you could meet a cuddle buddy, just as you would find a lover on Tinder. But companies like this had a tough time during the pandemic.

When we don't get body contact, it triggers a powerful stress response, a feeling of intense loneliness. Covid-19 made it very clear that we humans cannot cope solely with technological contact. We don't get what we need from a smartphone; it isn't warm, it doesn't smell, it doesn't breathe, our fingers slide over its cold surface without encountering any of the things that make human skin so interesting. Yet we touch it with our fingertips almost constantly, more than we touch our own children or partners, for example. Among other things, social media is a time thief. Regardless of the pandemic, we increasingly spend time being nonphysically present for each other, time we could, of course, spend attending to each other physically: being in the same room, laughing out loud together, looking into each other's eyes, or hugging—all ways of being together that seem to be vitally important to us. Even though the human brain has rapidly become used to being almost constantly in two realities at once (the physical world and the internet), it doesn't mean that the body no longer needs to be touched or seen or to just sit quietly with other people in the same room.

A large study carried out in Italy in 2020 showed that people who felt lonely used social media more often—which made these people feel more anxious and depressed, and thus even more lonely. According to the researchers from the universities of Enna, Naples, and Parma, this forms a clear negative spiral: you feel lonely, you turn to social media, which simply makes you lonelier because the contact you experience on social media isn't the contact you actually need. Researchers see that loneliness increased significantly among teenagers after 2011, the year that smartphone use among young people exploded. Since then, there has been an especially sharp increase in anxiety, depression, and loneliness among teenagers, as they increasingly spend time on social media. Another study showed a direct link between loneliness and social media: fifteen hundred teenagers who deactivated their Facebook accounts (today, it would probably be TikTok) were compared with fifteen hundred young people who continued using Facebook as normal. Those who deactivated their accounts reported spending more time with family and feeling happier, less anxious, and less lonely. This shows that while the nonphysical contact on social media can be good and important, it can never replace physical meetings with people we love.

Boston College philosophy professor Richard Kearney believes we are living through a crisis of touch. He claims that we live so much via screens that the body has been removed from our experience of life; there's been an "excarnation" of experience when it happens via the internet. Loneliness isn't just some abstract sense of not being understood or missing someone to laugh with, of not being seen, feeling trusted, or being included in a community—it is also something more basic. It is a feeling in the body. It is an awareness that there are no warm human bodies around yours.

In recent years, a lot of research has been done on internet use and cases of loneliness. Yes, young people report that social media makes them more connected to the world, and it's true that not being on social media at all increases the risk of lone-liness. But the research shows that you don't need to be online for many hours before it causes more loneliness than it does community.

There are several aspects of social media that drive loneliness. Posting something online and getting no response whatsoever is one of them. It is a clear show of disregard that feels like a rejection. If your social reality is played out as images and video clips on a screen, whatever happens in this sphere—what we only ever register with a glance—is hugely important. For the first time, there is a whole generation that has grown up with smartphones and the internet, and we don't know how being online so much really affects them. Nearly all Norwegian nine- to eighteen-year-olds have access to a computer, their own tablet, and their own cellphone—and make their digital debut aged eighteen months. Twenty-nine percent of the parents who were asked said that they always or nearly always sat with their children when they were online, but only 9 percent of these children corroborated this. So someone must be lying, and it's tempting to believe that it's the parents, because Norwegian children are number one in terms of being exposed to pornographic websites, bullying, and sending and receiving sexual messages. Fifteen- to sixteen-year-olds spend on average more than 3.5 hours a day online—nearly as much time as they spend meeting their friends in person. When not at school, about 30 percent of teenagers spend more than four hours a day in front of a screen. Teenagers turn to screen-based activities more often as they get older, and on a scale that has recently been growing. What children spend their screen time actually doing varies between genders. Boys spend more time playing games, while girls are more active on social media. During this time, the only contact they have is with a cold screen on which they can see extremely graphic content: 34 percent of Norwegian children have seen pornography, compared with 14 percent of European children, and 42 percent of Norwegian children have seen potentially harmful websites that publish hate speech, suicide-related material, and so-called pro-anorexia material. It worries me that children are drawn into algorithmic maelstroms of porn and violence, where they compare themselves with manipulated imagery; where they go hour after hour without body contact, without eye contact, smiles, or silence; where, instead of finding themselves and their own inner compass, they find a distorted reality. Social structures that encourage lies and masks create more loneliness. What happens when this replaces physical contact? Children need hugs, even more so than adults. Teenagers, including young boys, must have physical contact. Even young and grown men need love and closeness.

For several years, Lasse Josephsen lived in a parallel universe, a place devoid of any tactile experience. All online communities are built with your eyes and fingertips, but you don't touch other people, not really.

"I've always been drawn toward dark subjects," says Josephsen, who has spent years keeping track of the most disturbing parts of the internet. It became an obsession.

"I started every day by going on the internet. I found that any-one who talked about things that I wasn't interested in boring. It wasn't that I didn't have friends when I began; I just stopped being with them as much," he says today, about when he fell into the internet in 2007–2008 and stayed there. He wasn't new to the internet, of course, but this was the moment when it really began to occupy a big part of his life. The year 2007 was around the time that social media really took off, and that most people began using the internet differently.

For Lasse, it started with an attraction to death, destruction, negative emotions, black humor, and nihilism. And what was happening online was far more exciting than what he experi-enced in the real world.

"There were so many unpredictable things in my life, whereas on the forums I had some kind of control," he says. "I tried out various means of subversive expression and took control of the things I found scary. Yes, I've always felt slightly alone and an outsider in many ways, but being online like this is profoundly lonely.

"I wish we could get these men out of there. I know what it's like, and it's a deeply lonely existence. It was a huge relief when I realized that I didn't have to be in there anymore, that I could just stop," says Lasse, who now, many years later, has a healthier relationship with the internet and spends more time out in the physical world.

"I think loneliness is a serious condition that can do ugly things to people's psyches. When I've not had sex or a girlfriend for a long time, something feels off. And a little stroke on the neck from a girl's hand can be enough to make life worth living again."

The decisive step out of this murky online world was related to touch. A few days after the 2011 Norway attacks, Lasse closed his laptop and drove to Oslo. It would be his way out of the internet, the starting point, his way out of an isolated life in small-town Norway, which consisted mainly of virtual contact with Americans: Josephsen had met a girl on a forum for "negative people," and they were now going to meet in reality.

Their first meeting took place in a city ripped apart by a bomb, where armed police were patrolling the streets and thousands of mourners had laid roses outside the city's cathedral. Love can arise in the darkest of times, and the two became a couple and bought a dog together. Lasse loves dogs. Even now, though they are no longer a couple, he is still in touch with his ex and regularly looks after the dog, Arthur.

"Something happened to my view of humanity when I got the dog," he says. "It changed. I couldn't look at other people in a purely nihilistic way anymore. It helped viewing them a bit like dogs, as though a person I didn't like was a bit like an injured dog—that this person can be more than just an unpleasant obstacle in my daily life."

Animals have saved lots of people from loneliness. Therapy animals are used to treat traumatized children. Dogs used at institutions for the elderly and mentally ill significantly lower the residents' cortisol levels and reduce the levels of aggression and loneliness in people suffering from dementia. That's what the close warmth of a furry friend can do for us. When society shut down because of the coronavirus, there was a strong upswing in pet ownership in Norway.

"During the pandemic, we really learned how important touch is," says India Morrison, a professor of cognitive neuroscience at Linköping University.

"But I'm always touching this!" I say, holding up my smartphone, which is never far away.

"Of course, you touch it all the time, but it doesn't involve any of the complex contexts in which we encounter other people," Morrison replies.

Morrison's research reveals how complex touch is: It isn't one thing. How touch is perceived depends very much on the circumstances.

What's strange is how touch was the last of our senses to be investigated scientifically. Taste, smell, hearing, and sight have been studied for centuries. But touch, despite being perhaps the most important sense, has attracted the interest and excitement of researchers only in the last fifteen or twenty years. Research in this field has accelerated, however, and Swedish neurologists have developed tools for measuring touch—small radio transmitters for nerve fibers, almost, extremely small electrodes—which, when inserted into the skin, pick up tiny nerve signals. Microneurography has detected nerve receptors in the human body that only respond to light touch—C-tactile afferents.

"Touch is the most important of all the senses," says professor Francis McGlone. "Our research, and that of others, is showing that early life experiences of nurturing touch affect us in a wide range of ways, with a lack of touch during this critical time leading to a range of neurodevelopmental conditions such as autism and eating disorders. The knowledge of the importance of the role of C-tactile afferents in affective touch is hampered by the lack of neuroscientists skilled in the technique of microneurography, with probably only about twenty researchers worldwide trained in the technique. It is vital that this new knowledge about the importance of affective touch to our mental health (and we now know immune system) is communicated globally, as we are moving faster and faster into a 'untact' world."

McGlone began his own research career by studying skin pain and itching, temperature variations and touch. But in 1998, after moving on from academia to work in a laboratory investigating skin care and our use of skin-care products, he happened to read an article about C-tactile afferents by the Swedish researcher Åke Vallbo. Stumbling upon this research paper would change McGlone's life and make him a leading expert on touch, although he didn't know it at the time.

We now know that our skin contains three types of thin nerve fibers: those that register pain, those that register itchiness, and those that register stroking, each of which, in their own way, help us encounter the world. They both protect us and are related to pleasure. We've long known that there are nerve fibers in our skin that sense danger, that signal to the brain that we've been pricked by a thorn or burned by the cooker. These nerves are encased in a layer of fat called myelin that isolates them, allowing the electrical signal to move extremely fast, just a few tens of milliseconds, from the painful area to the brain. This is something we very much need when we are in danger. It makes you quickly pull your hand out of the fire or the thorn bush, before we notice the pain a fraction of a second later. The pain signal is sent via thin nerve fibers that are not encased in myelin, slowing the signal, which takes a thousand milliseconds to arrive. And hidden throughout our skin, where there is and has been hair, are tiny nerve fibers tuned to an entirely different kind of touch: cuddling. It was Åke Vallbo who discovered these nerve fibers, but there are now researchers in Linköping, Gothenburg, and Marseille—as well as in England and the US, of course—all studying how these nerves function.

For a long time, C-tactile afferents were a mystery to scientists. But it's possible that these relatively slow nerves are extremely important for our sociality. For our identity. For our happiness. These nerve fibers are connected with loneliness.

"These C-tactile afferents are the key to understanding our social life," says Francis McGlone. "The slow nerve fibers help us understand how we relate to each other, how we develop an identity, and what happiness is for us."

McGlone sees the world from the viewpoint of touch: What happens to prematurely born children who have been kept in incubators? They are babies, and they are inside transparent boxes where they can only be touched in exceptional cases.

"I wanted to locate babies that didn't experience much cuddling during the first months of their lives, as was found in the 1990s in Ceauşescu's Romania, where thousands of babies were brought up in orphanages without adequate nurturing care, leading to most of them displaying severe behavioral problems. I wanted to investigate the impact of this lack of nurturing touch in early life, which is when I realized that preterm babies, when placed in incubators, would no longer be getting the touch they were receiving when in the womb. As many as 25 percent of these children develop autism spectrum disorders. We think that it is because of the lack of touch."

McGlone's latest research project is examining the very same incubator children. Touch is crucial to brain development, for the immune system, for growth. Experiments with the tiny roundworm species *C. elegans*, which cluster upon hatching, showed that individuals that are separated from the cluster grew significantly less than the others. They remained tiny, compared with their hypersocial siblings—their growth was somehow stunted.

The ideal stroke is even possible to measure: when you stroke a baby, it will instinctively relax when your hand moves at a speed of three to five centimeters per second. I'm not joking! Any faster or slower and the nerve fibers won't detect it. Researchers assume that the ideal stroke provides optimal conditions for the secretion of attachment hormones such as oxytocin and endorphins in the brain.

In an experiment conducted by McGlone and his research team, one group of rats was stroked for ten minutes at the perfect speed: five centimeters per second. Another group of rats was taken out of their cage and given ten minutes of very stressful stroking: at thirty centimeters per second. Both groups of rats were then exposed to a "chronic mild stress" program for two weeks, where their ability to handle stress was measured. During this test, the rats that had been stroked incorrectly and too quickly became panic-stricken, while the rats that had been stroked properly coped with the stress. Stroking is connected to stress management and emotion regulation. Baby rats won't be well fed or cared for by a stressed and hassled mother rat, making them far worse at handling stress than adult rats, compared with rats that were stroked properly and frequently. The study also showed that the female rat pups that were underfed when young became equally bad mothers to their own rat pups, thus proving what longitudinal studies of humans like the Minnesota study showed: That we learn our behavior from our parents. That stress is inherited. Our early experiences with touch can affect our ability to care for others in the future.

And stroking also relieves pain: Professor McGlone did a large study to see how babies handled the pain of an obligatory blood test, taken from the underside of the child's foot to check for diseases immediately after birth. In each case, the prick of the needle was clearly painful, and the researchers were able to measure the pain response via electrodes attached to the babies' heads— as well as by the screams, of course. However, if the babies were stroked perfectly—at three to five centimeters per second—just before getting the jab, the pain response stopped almost immediately. If they were stroked faster, the pain continued.

But why am I talking about rats and babies and right and wrong types of stroking when this is a book about loneliness?

"One of the most important roles for C-tactile afferents is that they measure loneliness. Going without physical contact with others, as you do when you're lonely, actually leads to a shorter life," says Professor McGlone.

C-tactile afferents cover the entire body, including your back, which you cannot reach yourself. And what researchers now know is that stimulating the C-tactile afferents lowers the heart rate and ensures the secretion of serotonin. When we experience a positive touch, we feel a sense of belonging and calm. The touch between a mother and newborn child also helps make connections between the child's brain and body; it allows us to know where our body is. It helps us to understand where our own perimeters are. What the research on C-tactile afferents has shown is that we need touch for security, for growth and brain development, for regulating emotions and calming the nervous system. The problem is that touch is extremely dependent on context.

"Obviously, there's a big difference between being stroked on the thigh by a scary Harvey Weinstein type and being hugged by your boyfriend," says Guro Engvig Løseth.

Løseth is a PhD candidate in psychology at the University of Oslo and an expert on brushes—fine-bristled goat-hair brushes, which she uses in her research on touch. By conducting an experiment where the test subjects were stroked on the arm at three centimeters per second by someone they couldn't see, after taking drugs that either increased or blocked the release of endorphins in the brain, Løseth aimed to find out if endorphins are what actually make stroking so nice. And as bizarre as it was to be stroked with a goat-hair brush by a stranger they couldn't see, people responded positively. However, Løseth saw no effect from the drugs—the subjects reported liking the experience even when nearly all the endorphin receptors in the brain had been blocked. This was quite a different outcome from the results of the experiments on monkeys, though in the case of the monkeys it was often their friends who did the touching. To control her experiment, Løseth had removed the social context from the touching: it is the context that determines the meaning of the touch. This makes touch difficult to research. Measuring visual impressions, for example, is far easier, because measuring how a subject reacts to a set of images is a bit clearer. But when we are stroked, something mysterious happens, something that involves more than just oxytocin and serotonin. The human brain is not a machine. It is governed by associations and emotions and context. It is relational. And perhaps even more so when it concerns the most intimate of senses.

When I meet Løseth to talk about touch, she is heavily pregnant with her second child.

"Has what you know about the research on touch changed anything about how you want to raise your children?" I ask.

"It's made me even more conscious of how important hugs are for children," she replies. "After what I've learned about the importance of touch, there's no chance of me not co-sleeping with my kids. No other animal parent takes the most vulnerable member of the pack and moves it far away! Of course children should sleep beside their parents! It's become very clear to me that we have to cuddle a lot, and sleep next to each other, close to our children, if we can."

One of the most important functions of touch is to give us a sense of closeness with our caregivers. It enables us to feel their warmth and feel safe. Imagine kittens all snuggled up with their mother—the warmth in that pile of animals is essential to their survival, just as it is to human babies. Ideally, every human should be wrapped in several blankets that keep them warm: blankets of love, understanding, trust, and comfort. Perhaps this is what frightened little Toffle was trying to achieve when he turned on all the lights in the evening and snuggled under his duvet and blankets; it's perhaps why Aureliano Buendía goes around wrapped in a wool blanket in the heat of the Colombian summer: to get this feeling of being warm, protected, and comforted—a feeling that only humans can really give us.

"The C-tactile afferents in our skin respond best to being stroked by humans and respond most to something that is 32 degrees, which is the temperature of human skin," says India Morrison, at Linköping University. "This means we can perceive these nerves as having been constructed solely to detect us being stroked by other people."

But there's still a lot of mystery around the importance of touch.

"We call the stimulation of C-tactile afferents 'social touch.' But there are many things about this type of touch that we still can't quite get to grips with," Morrison continues.

One of Morrison's recent experiments shows that if we are stroked by someone we know just before a stranger does the same thing, we become calmer and more secure from being stroked by the stranger than if the sequence was reversed and the stranger stroked us first. But the effects were not significantly different.

"People are incredibly different as well—we don't even know what a normal need for touch actually is," Morrison says. "We don't know how it varies! And it's possible that the map we draw of social touch is largely governed by the measuring instruments we use, those that measure oxytocin and C-tactile afferents. We also can't say why it's so nice to be touched on the palm of the hand—we've never had hair, nor do we have C-tactile afferents, on the palms of our hands—and yet this is the most sensitive part of our skin; people report it being just as nice to be touched there as it is on an area that does have C-tactile afferents. And we also don't know why C-tactile afferents are linked to the skin areas that do have hair follicles."

For Francis McGlone's and India Morrison's research, the pandemic will have been a kind of gold mine: there is much to learn from the world's population being prevented from touching each other. What happens to us when all the little touches that human life is full of are gone? During the pandemic, all these touches—the consolatory pat on the back of a troubled friend, the extra-long hug for a recently married woman, the comforting grip on the hand of a frightened old man, the gentle nudge of fellow passenger on the bus, the breath and warmth of a stranger sitting on the neighboring seat in a dark cinema— were taken away from us. What did this do to us, what did it do to our societies? None of us can know for sure. What we do know is that loneliness has skyrocketed.

Tiffany Field, a professor of pediatrics, psychology, and psychiatry at the University of Miami, found that the amount of touch experienced by teenagers had a crucial effect on their aggression levels. Field had observed a clear difference between American and French teenagers; the French teenagers had far more skin contact and were less aggressive and combative. When the American teenagers then received a daily massage from a family member, over a period of five days, their aggression levels decreased noticeably. In April 2020, American families were examined by researchers, who found that only one-fifth regularly kissed or touched their children. Touch calms us down.

"Being overlooked, rejected, invisibilized, and not touched, as in, totally aside from violence and abuse, creates a strong feeling of loneliness," says trauma expert Mari Bræin. "Babies will already show signs of stress in the first months if their parents are dismissive and unavailable. They are so eager to make contact, they'll try numerous ways of getting their parents to react. But the parents might then react by withdrawing further, perceiving the child to be clingy, which makes the child even more frustrated."

Children who experience this have been followed through so-called longitudinal studies, conducted by Harvard professor Karlen Lyons-Ruth and her research group, who have found a connection between this type of rejection and problems that these children develop later in life—problems socially, problems with emotion regulation, and suicidal behavior. Video recordings taken by the researchers when the children were aged four show how the children take responsibility for their parents by praising, maintaining, and regulating the adults' emotions. The interaction was marked by role confusion, where the child would, for example, bend over a drawing the mother had done and compliment it, as a parent will often do with their child. But all this commitment to getting a response, to getting love and physical closeness, can do something to the child's sense of self-worth.

"The children become hypersensitive to other people's needs, but learn to disregard their own," Mari Bræin says. "They learn that it doesn't matter what they feel and need, they don't even know what they feel, because they haven't learned it. It is a great loneliness. The study also showed that many of the children took their own lives before they turned thirty."

Professor of psychiatry Bruce Perry is considered an authority on trauma treatment and works to translate brain research into practical treatment of traumatized children, as head of the ChildTrauma Academy in Houston, Texas. Perry's research and methods are among those used by Mari Bræin when she works with children and young people who are exposed to neglect.

"When rational behavior is not rewarded, the brain finds other ways of reducing discomfort," Bræin says. "Many children who have been too often alone at the orphanage and not been given the care they needed would rather be given candy than hugs and human contact when they are in pain. The child's brain will have created an association between sugar and pain relief, when relational support has, for them, been either absent or unpredictable. They never know how long a relationship will last, or if it's a good or bad one, whereas sugar is a more reliable source of comfort. And this creates intense loneliness."

Children who experience insecure attachment—indeed, where the adults in their lives are simply guests, and they receive very little safe touch—find other ways of easing their discomfort. The most appropriate thing, of course, is to get closeness, comfort, and understanding from someone who loves you. But stimulation like sugar, salty and fatty food, drugs, and sex also work effectively in the reward system. They are short-term cures for uneasiness and anxiety.

"All these other forms of stimulation can ease the pain there and then but can lead to addiction and health risks in the long term," says Bræin. "Loneliness, and the sense of not feeling loved, is repeated in all the different forms of neglect and stress I have seen. It also applies to sexual abuse and violence and emotional neglect. It's a common theme. If no one even helped you learn what your body needs and who you are, how are you supposed to handle being in close, long-term relationships? You lack the most basic relationship skills, which means you will continue to seek destructive comforts and destructive relationships."

A child's brain will grow and develop much like a blooming flower, provided the child is given all the love and secure closeness it needs—first the brain stem, then the cerebellum, the limbic system, and finally the advanced and higher functions of the frontal and temporal lobes. But if the brain's basic ability to regulate emotions—which is the first step toward becoming an adult—doesn't develop, it will inhibit the child's remaining development. Being regulated means being able to understand and manage one's own needs and feelings, starting with the very first touches a baby receives, and continuing with every subsequent touch, each one wrapping a thin, protective blanket around the child, giving it resilience, a powerful ability to survive—like an emotional swaddling of the child. Trying to do this when the child is older is more difficult, though not impossible, but it has to be an entirely physical process: it's not something a therapist and client can accomplish simply by talking.

"Stimulating deeper brain structures through physical activity and sensory work is believed to strengthen the neural networks that regulate stress responses," writes Mari Bræin in an article cowritten with her research colleagues Heine Steinkopf and Dag Ø. Nordanger. In their therapeutic practice, they use music, rhythm games, swimming, climbing, massage, yoga, and playground swings; each child can lie in a hammock, use a balance board, run an obstacle course, jump on a trampoline, or do other jumping or hopping activities. At school, the children are given breaks from lessons to engage in sensorimotor activities such as standing up and stretching, clapping games, or similar things. It is a therapy that uses numerous ways of reaching the child's sense of play and physical security. Therapy animals also have a very positive effect on children suffering from this kind of damage; a calm, breathing, living creature can be cuddled or ridden without language. Children can also be wrapped in heavy blankets or use a so-called body sock, which gives them a sense of being held without involving another person. This helps children reconnect with their bodies and makes self-regulation easier for them.

"To create change, we have to go deeper into the brain," Bræin explains. "In the absence of words, we have to work on establishing a bodily experience of security, which often requires more than just talk therapy."

Researchers believe that not being comforted or emotionally regulated by an adult at such a young age will affect a child's entire range of functions, as far as and including abstract thinking, consequence assessment, concentration, focusing, and ability to find direction and set long-term goals. This renders the child totally unprotected when encountering the world and can result in a child who overeats or is unable to sit still or bangs their head against the wall at the tiniest amount of stress, much like children who are wrongly diagnosed with ADHD. These children become self-stimulating, like the children in the Romanian orphanages who rocked back and forth or just stared at their own hands.

"We often think that love is the answer," Bræin says. "But it's a bit more complicated than that, because for some of these kids, the scariest thing of all is closeness. That's when they put the spikes out."

One ten-year-old boy Bræin worked with struggled to accept cuddles or physical contact from his foster parents. At the same time, he wanted them to help him with numerous things he was actually capable of doing himself, such as tying his shoelaces. Just like the Groke and the hurricane lamp, he needed to find a detour to receiving care.

"I advised his new caregivers to be especially thorough when tying those shoelaces, because he could only handle a very indirect form of closeness," Bræin says. "For many children, closeness and cuddling can be too sudden and suffocating. Many of them push their new caregivers away, which makes the foster parents feel powerless and give up. So the very thing the child expects and fears actually ends up happening: he is sent back to the institution. And the institution might well feel easier, because there isn't such a close relationship there."

The children Bræin works with have a basic expectation in life: that it's *only them*, that they are totally alone. It is a physical experience of loneliness.

"For every single person I've met through child services and acute psychiatry, loneliness has been a central theme in their lives. So how can we find a closeness that they can handle? You can expand the intimacy barrier, but that often takes a long time and can be very painful for all involved.

"The great thing about humans is that we are plastic. There is reason to be optimistic. Because even people who have experienced violence or lots of negative things will be able to make changes. The key is new relational experiences, hard work, and reflection."

A totally new method is now being used to tackle the profound loneliness many of these children go on to experience as adults. Very little research is available on these adults—who often end up committed long-term to psychiatric institutions—because so few treatment measures really work, which in turn means there are few reportable figures and results. These people are often also heavily medicated, enveloped in labels and diagnoses; everyone has given up on them, and they have given up on themselves.

Now, however, they can be sent to Didrik Heggdal, a specialist in clinical psychology whose treatment is based on exposure therapy. Exposure therapy can be used when, for example, you are afraid of spiders: you begin by looking at a picture of a spider and gradually work your way up to being fully exposed to a real spider, until your tolerance has been sufficiently built up. Heggdal's treatment, however, exposes you to *loneliness*.

"The people we treat are afraid of affective activation," he says. "For them, it's scary to feel anything at all, but many of these people even try to avoid the experience of pleasure, because it's a feeling that's on a par with anger and sadness, for example, which makes you potentially vulnerable to a lot of pain."

All feelings are dangerous when your experience of loneliness is so physical, because all feelings are linked to relationships, to other people, both happiness and sorrow. Being numb like this is lonely, because you are lonely when you are a stranger to yourself.

"The idea behind basal exposure therapy is based on existentialism and not least on a fundamental belief in people's ability to regulate themselves. The patients we treat here have locked themselves into a severe phobic state with an existential fear of catastrophe," Heggdal says.

In his book *Berøring* [Touch], psychologist Peder Kjøs investigates everything from sex dolls to the kind of emotional scars a violent mother can leave on her son to the need to give a client a hug. A therapist can't really do that. "Skin contact is, of course, our primary way of regulating emotions," he writes, "it's our first encounter with the world, our first means of communication with others and of experiencing rejection, it's our sense of being cared for and feeling safe. All these things happen exclusively through the skin at first, before hopefully striking a balance between too much and too little skin contact as life goes on."

Kjøs finds it hard to sit with a client and not comfort them with a hug. In many cases, it will seem obvious that it's what they need, in addition to the talking. "In situations like this, there are fine lines between what's okay and what's not, but we are born with very different needs for touch," Kjøs says. "We don't have much of a culture for touching in Norway, and problems can also occur with clients who have experienced violence or sexual abuse. It affects how we experience touch."

So touch is ambiguous, multifaceted, and extremely important to us, and at the same time a potential minefield of misunderstandings and #MeToo accusations. Because what's an assault if it isn't forced touching? Sexual assault means that someone has forced themselves into your intimate zone and done something to you that should normally feel good and should ideally be an expression of closeness or at least trust. And this is where rape and loneliness overlap. When you are raped, you are an object in another person's hands and feel robbed of any care or protection—and in many cases, you know that you're in danger of being killed. Rape only looks like sex, and it preys on all our sexual feelings, but it is pure violence.

Rapists are getting younger; nowadays, they can be boys as young as eleven years old. Children have become used to porn: the average age for seeing online porn for the first time is eight. Young women are choosing to be celibate rather than have so-called porn sex. Amia Srinivasan, a philosophy professor at Oxford University, writes about how sexual liberation has failed the young, who are now being trained in sexual puppetry via the internet: "It etches deep grooves in the psyche, forming power-ful associations between arousal and selected stimuli," she notes.

Watching porn can shape the human gaze and our sexuality into something mechanical and strange. When we think that intimacy should happen in a certain way, how does that affect the way we touch each other? In many ways, to reenact porn scenes with your lover is to sexually objectify yourself. We do it to ourselves, like a form of violence we inflict on our own bodies and our own physical needs.

"What we need isn't a kind of positive hermeneutics to be inculcated in viewers of pornography," Srinivasan says. "What we need is the onslaught of images to just *stop* for a moment."

New research from the Norwegian Center for Violence and Traumatic Stress Studies showed that 22 percent of women and 3 percent of men had been sexually assaulted, abused, or raped. Half of the women raped were raped before they were eighteen. Most rapes and attempted rapes also occurred in an existing rela-tionship, and very few of them were inflicted brutally, as in the case of aggravated rape.

"'Women actually like being raped.' 'Any woman who *really* wants to can avoid being raped.' Rape myths like these construct an ideal of victimhood that is as unattainable as it is desirable. They create 'worthy' and 'unworthy' victims," write rape researchers Anja Emilie Kruse and Anne Bitsch in the book *Bak lukkede dører* [Behind closed doors].

But it's not just the rape myths, the shame, and the self-loathing that affect you after a rape. Just as positive touch creates good connections between the brain and the body, negatively experi-enced touch—from violence or sexual violence, being touched in an undesirable way—can seriously disrupt a person's relation-ship with their body. All touch depends on the context: who is actually touching you and why? Touch can create an unparalleled feeling of security and closeness, and it can be the most powerful symbol of distance, of loneliness. When we receive a loving touch, it can initiate attachment and calm, or be a trigger for memories of violence and rape.

A sexual assault can leave you with a strange relationship with your body, which affects your own physical experience of yourself. It turns your body into a crime scene, full of evidence and reminders, and you feel like some of your immediate access to who you are has been lost. You become a stranger to yourself and your own needs, because you have been used as an object by another human being. In a large study involving sixteen hundred participants—half of whom were survivors of abuse, the other half a control group—rape was found to affect self-image, body image, and sexual relations and found to cause depression and anxiety. What research on rape and trauma also shows is that being raped will often send you into an extreme stress reaction, PTSD.

It was, of course, here that Vincent Felitti's research began, with his obese patients telling him about the abuse they had experienced. A rape transforms you from an individual with your own value in a community into an object, someone unworthy of love and protection. Being raped is pure loneliness. In *Bak lukkede dører* [Behind closed doors], Anne Bitsch and Anja Emilie Kruse describe how it feels to have suffered rape: "When I carefully shut the door behind me and go out into the autumn night, I feel like the loneliest person in the whole world. But I'm no fucking victim."

An abused person can be enveloped by layer upon layer of shame, which affects their relationships with other people: they won't trust people as they once did, their trust will have been fundamentally shaken, and they will no longer be able to separate those who mean them well from those who don't.

"He never had known whom to trust: he had followed anyone who had shown him any kindness. After, though, he decided that he would change this. No longer would he trust people so quickly. No longer would he have sex. No longer would he expect to be saved." This quote is from the story of Jude in Hanya Yanagihara's novel *A Little Life*, about a man who has grown up suffering extensive sexual abuse by a Catholic monk, as well as violence and mistreatment. Jude becomes a successful lawyer and gets a boyfriend, Willem. But at night, he self-harms, and his arms are covered in cuts from his razor blade and burns from the oil he lights the fire with.

Self-harming is a way of dealing with painful feelings, a form of coping strategy that isn't actually coping but a replacement behavior. It is the body speaking, not rationality, because most people engaging in self-harm know that it only has a short-term effect. It eventually eats its way into every relationship, becoming a coping strategy for more and more areas of that person's life. Those who engage in self-harm often cut themselves increasingly deeply, which can then become an involuntary suicide attempt. This isn't the intention, as a rule, though the cutting is an expression of strong inner pain and rebellion. The pain from the lacerated skin is a distraction from the internal pain. It can also be a way of indirectly signaling to others that you want to be cared for, like the little boy who struggled unnecessarily with his shoelaces.

Martin Eia-Revheim was such a child, cutting himself as a way of dealing with his violent father: violence born from violence. And in a way it helped, when he was totally alone and had no other options: "As the blood hit the basin, the pressure subsided," he writes in his book *Å sette sammen bitene* [Putting the pieces together]. "Not that easing the pressure through blood-letting would resolve everything, but I still think today that it offered some kind of release, that all the things that were unclear, dark and chaotic, became real, turned red, and ran slowly out so that I could wash it all down the plughole. I don't remember what I was thinking the first time I cut myself as a twelve-year-old. I must have known that it helped, that my head became clearer and my breathing calmer, because I carried on doing it whenever I was unable to regulate my feelings another way."

Britain's National Health Service, the NHS, has several recommendations for people who cut themselves. The first piece of advice, "Talk to a friend," is perhaps the most difficult, because the shame and loneliness associated with self-harming make it worse.

"The weird thing about shame," says shame expert and researcher Helene Flood Aakvaag, "is that it makes it easier for you to be revictimized, to experience something similar again. It's a mystery to us, but if you've experienced something traumatic and are ashamed of it, there's a far greater chance that you'll experience the same thing again. If you experienced violence as a child, or were raped, it's more likely that you'll experience violence and rape again if you're ashamed about it. We don't really know why this is, but it could be that if you're ashamed of something, you don't talk about it with other people. Whereas if you share these things, it can encourage those around you to protect you, as well as the fact that if you're unashamed, it's more likely you'll be drawn toward good relationships."

I see myself in all this: the quiet girl standing alone at playtime or sitting by herself with a book while the other children play tag. The little girl sitting on the lap of random women, who takes the hand of strange adults, who goes home with the wrong man. I see myself helpless on a bed, a girl who has vacated her own body. My skin has been a lonely landscape. But it's not something I can draw a map of. I know that I don't understand my own body: I have no sense of my own needs, of whether I'm hungry, full, tired, or happy. I can mistake anger for hunger and pleasure for fear. But what I now do more often is ask my husband for a hug. And I bury my nose more often in our cat's silver-gray fur and just stay like that, totally calm. I don't really have to look at research to know the importance of touch, because getting too little of the right and too much of the wrong kind of touch has shaped my life. My daughter, however, can sit on my lap and get hugs and kisses whenever she wants, sometimes a bit too much—"Stop it, Mom! Enough hugging!" she'll say teasingly—and I feel confident that there's at least no lack of physical contact in the home I have created. Perhaps my loneliness began in the incubator I was placed in as a baby? My life began in 1975 in a plastic box, while my daughter began hers snuggled up to me, and so on. I knew she needed to be as close to me as possible.

The mosts intense form of touch, however, is the one we experience with other adults. Negotiating the intimacy of love and sex can be extremely difficult. We can never reexperience the intense vulnerability of a child, other than in rare, fleeting moments. But we perhaps experience a similar interplay, between closeness and distance, when we are alone, skin-to-skin, with another adult.

"There's something ambivalent about an intimate relationship," says Peder Kjøs. "You can't merge with the other person entirely, no matter how much you want to. There's always an element of tension between trust and rejection."

I met someone who looked into my eyes, who made me laugh, who wanted to understand me (even if it wasn't always possible), who held me, who stood by me when I needed it. Magic can sometimes happen, when we experience this most extreme and intimate form of touch. It happened to me: I know it shouldn't have been possible, but I knew it long before I took the test. I had become pregnant. A tiny little life was clinging to my womb and being squeezed by it. One touch had led to another touch, as though it had been passed down by thousands of generations. Some of my foremothers had more than likely been raped, forced to bear children. But some of them were hopefully like I was: a lovestruck expectant mother with a shooting star in her belly.

I felt her leg kick at the wall of my tight and large stomach as I swam in the turquoise water of a Greek bay, and it felt strange to be touched like that, from the inside. Floating in the sea, with a child floating in me. And at night I would stroke my stomach and feel her knee digging against my ribs. This life-giving touch had occurred again.

8

I encounter
the Groke for
the last time

I WAS WOKEN by a loud and cold-sounding bang. At first, I
thought it was thunder, though not the usual rolling rumble
that spreads across the sky. This was more metallic, short, and
clipped. I had been sitting in my apartment in Oslo, writing a
novel about loneliness and our desperate yearning for love and
community, and had then lain down on the sofa for a short nap.

But it wasn't thunder. It was a bomb.

Moments later, I got a call from Ada, who sounded distressed.
She had just walked through the foyer of the Government Quarter's main tower block, minutes before it became the epicenter of
the blast. The seventeen-story building, called Høyblokka, was
now just a shell with broken windows. Images soon fluttered
across my TV screen, showing downtown Oslo looking like a
war zone. Where did this evil come from? And why didn't I realize what was going to happen to Ada until it was too late?

One June day in 2013, while scrubbing the floor in my apartment, I found a strand of hair, long and black and thick, which had snuck in between two floorboards. I didn't understand where it had come from at first, just put it on the dining table and looked at it until I suddenly realized whose it was. It had to be Ada's. As if she had just dropped by and left a business card. By then, it was many months since she'd gone. I didn't realize it at the time, but this strand of hair and the explosion would help me draw a map of loneliness. Ada's loneliness would help me better understand, her solitary hair in my hand pointing the way in the darkness; it would be the thread of my story, my strange-looking compass needle guiding me as I ventured to my map's South Pole, to the frozen wasteland in the center—where Ada found herself in the cold.

I didn't know her that well, but I imagine she was both scared and full of expectation when she came to Norway from South Korea, aged four. The people she would call parents and who had been given responsibility for her were seemingly resourceful people with children already and well equipped to look after this little girl from the orphanage. You might say they were pillars of the community, because they contributed to the neighborhood, they had good jobs, they lived in a medium-sized city in northern Norway. It must have looked good all-round on paper. Adopting in Norway is quite difficult, so they would have been carefully checked. The case handlers would have known this was a child who needed someone who could give her all the care in the world.

I only know what she told me: that in Norway she had to learn the language and culture and find her place in a dysfunctional family. It was to be an upbringing marked by neglect, where she was considered less worthy than her siblings, the family's biological children. She described feeling like a servant in her own home. She told me about her parents' alcohol abuse. She didn't feel safe or feel like she had a home. She didn't feel loved. Being adopted when you're as old as four is, of course, especially challenging. And was she really four? It seems very unlikely that you can be sent to Norway, from South Korea, at that age, with language and memories and thus a sense of losing everything you're familiar with! But I check and find that, yes, it has happened. And if you are four at the time of adoption, your new parents will have missed all the decisive moments of connection. All your first smiles and looks, your first touch, will have been with other people; your new parents weren't giving you trust and security from the very beginning. If you have lived in an orphanage and never had any kind of permanent caregiver, bonding well with only two parents is hard enough. But what if the people who become your caregivers here in Norway aren't suitable? What if a child, longing to be loved but speaking an entirely different language, who has come all the way from Asia to find safety, isn't seen or understood or wrapped in multiple blankets of love? What then?

From an early age Ada was interested in singing and sang in several choirs as a child. And when she and I became friends, she was bursting with life, with a glowing smile and thick, flowing hair. We went to concerts together, because her love of music never faded, and I remember her eyes, which were full of life and curiosity. And I remember hugging her, which was like picking up a long-haired cat and realizing how skinny it is beneath its long, shaggy mane.

She must have felt some kind of certainty that her closest caregivers didn't love her, and this bled into all her subsequent relationships. Romances came and went, but she was never able to make them last long enough to start her own family, no matter how much she wanted to, I know that. She had a lot of friends, but few of them knew about her dark secrets—she kept up a mask. She was also known to be great at her job. I think she wanted to help other people and repair all the problems she couldn't repair in her own life.

That was Ada, maybe: someone who wanted to make the world more accommodating, though I can't be certain what she was driven by. But I can assume that she probably experienced racism, a form of racism so veiled that it goes barely detected: racism against foreign adoptees. One reason for this is presumably that those adopted from abroad are officially considered fully Norwegian, and in all the registers and paperwork, that's how they are listed. So the parents of those adopted from abroad are not prepared for the fact that their children might suffer racism and are unable to cope. It is this type of racism that artist Uma Feed, who was adopted from South Korea, has experienced a great deal of. And what she has been exposed to since her early teens has been explicitly sexualized racism, usually from older men.

"It's the purest form of racism, because it's solely about my appearance, my ethnic genetic features," she says. "The language, the culture, and all the traditions I relate to are totally Norwegian."

Feed has written and directed a film about what being in this kind of situation is like. Her loneliness has been compounded by the fact that, until now, no one has really acknowledged that this racism even exists. When she visited South Korea, she felt equally lonely: she wasn't Korean. Inside, she was totally Norwegian—she didn't understand the language and didn't know the culture. Nevertheless, for the first time in her life she was able to walk down the street and feel like she was part of the majority.

"I live with a double loneliness," she says.

She believes that the experience of being left out on a street or on the steps of an orphanage is so embedded in her body that it has left her with a permanent sense of abandonment and feeling excluded. She was a vulnerable child left totally unprotected and ready to die.

"I probably have a special relationship with death," Feed says. "I accepted it once and for all and became nonchalant about life. And I don't connect with people like others do. Children in orphanages get attachment disorders and run to anyone that might offer them care, which is, of course, the best survival strategy in an orphanage. So you run in the opposite direction to your new Norwegian mother, you run to whoever is offering you love, because you can't be attached to just a limited number of people. I feel like Batman: I'm boundlessly independent and the spikes come out if someone wants to take care of me. But I'm also afraid of losing people, of falling in love with someone and losing them again."

Feed is troubled by a deep sense of loneliness and abandonment, and at the same time she cannot get an answer as to where she comes from. People also constantly expect her to be grateful.

"I hear it regularly. That if I'd lived in South Korea I would have been poor or a prostitute—that I should be grateful. But what do they know?"

Feed is critical of the entire foreign adoption system, which has been investigated and stopped in countries other than Norway. There are plenty of reasons for this. First, the UN has decided that there should be a global discontinuation of these types of adoption. It has also come to light that often the parents in the country of origin have only put their child in the temporary care of an orphanage and later return to find that the child has been given up for adoption. Some children are blatantly stolen from the streets outside their homes. Feed believes that countries like South Korea have been treating children as an export commodity.

Now, with the war in Ukraine, children are being taken from there, just as they were taken during the Korean War. In war, children are the first to suffer.

"There will always be rape, war, and children born out of wedlock, always a market for selling children," says Feed. "So I'm not against adoption, but the way it's been handled so far has been on the verge of human trafficking."

According to a Swedish report, internationally adopted children are near the very top of the national suicide statistics. Norway has no similar overview, but a report from the Institute of Public Health states that, as a group, "internationally adopted people have a higher incidence of mental disorders and, among other things, behavioral difficulties compared with the overall population. Both nationally and internationally adopted people have more frequent contact with mental health services than the overall population. It is also found that, in terms of suicide and alcohol-related causes of death, both nationally and internationally adopted people have increased mortality."

The Swedish report showed that, on average, internationally adopted women attempted suicide 4.6 times more often than any other group, which is startling because suicide is most frequent among young men. In the last year, as a result of Black Lives Matter, a campaign to fight racism against Asian-born people has also arisen in Norway.

For many reasons, after the July 22 attacks, Ada's life became lonelier than ever, and she withdrew from me, perhaps not intentionally, but that's how it turned out in practice. She had to travel more, she was often living in Bergen because she found herself a boyfriend there, so I didn't see her as often anymore. Did she get PTSD? Was she woken up at night by dark nightmares, was she startled by loud noises, was she afraid of large crowds and inner-city areas? I don't know. As I remember, she became a shadow who just slipped away.

But she came along to the party when I launched my first book, the one about longing and loneliness that I'd been writing the day the bomb went off. It had taken me a year and a half but was finally finished, and the big party I threw marked the day that I finally became visible. That's how it felt.

Ada, however, looked haggard and pale, and she didn't smile properly. There seemed to be a shadow across her face, but no more than there usually is when you've slept badly or you're hungover.

"I didn't bring any flowers," she said apologetically when I greeted her as she entered the venue.

"That doesn't matter," I said, and in hindsight I'm glad that's what I said to her. "You didn't need to bring flowers. You *are* a flower. I'm just glad you came," I said before hugging her, and beneath all the hair she felt skinnier than ever. Had I known what she'd been through a few months earlier, I would have been amazed she was there at all. But I didn't know. She hadn't told me about her spiraling depression or about the doctor who sent her home despite her noticeable injuries after a suicide attempt, telling her they'd be in touch, leaving her alone with her thoughts as she waited for them to call. But they didn't call.

As I read aloud from my book, Ada watched me from the doorway with her arms folded. She was the only person who stood like that, as though she was about to leave.

But how can I really tell this story when its main character is no longer here? What do I know about her inner life, what do I know about the inner lives of any of my fellow humans? What traces do we leave that might tell the world how lonely we've been?

The day after my book launch, Ada killed herself. She, who on July 22, 2011, had walked right past a one-ton bomb and survived—yes, she who had so clearly cheated death—no longer wanted life.

To be continually reminded that you look different is intense loneliness. Ada never mentioned the racism, because she was distinctly concerned with looking strong. Being a child who covers up for your own neglectful parents—that is loneliness, knowing that your guardians don't consider you worthy of protection and understanding, knowing that you don't belong anywhere. I often think about the scars she bore, the pain she had to deal with on her own. I don't know if she was traumatized by the July 22 attacks; she seemed so tough on the outside. All the weak points in her life were hidden away. She shone so brightly.

I struggle to imagine what it must have been like for her at the very end. Was it the power and determination she had in life that made her succeed in taking her own life? Could she have been saved? Would she still be alive if the health care services had immediately hospitalized her following her suicide attempt?

At Ada's funeral, I was both distraught and furious. It had happened so suddenly, and when I was at such a completely different place in life than she was. I wish I could have prevented it. I wish I had understood. Looking back, with all the facts at hand, it seems obvious that I should have tried to intervene. But at the time, she had withdrawn and vanished into her own pain. At the time, it seemed so difficult to help.

TV presenter and comedian Else Kåss Furuseth experienced the suicides of two close family members. The first real sign of trouble was perhaps in the autumn of 1991, when her mother—a normally funny and unpredictable, well-groomed and smiley person, who worked as a doctor—was admitted to a psychiatric ward. But Else thinks it may have begun sooner; she remembers something being wrong much earlier that year, when Norway's king, Olav V, died. It was January, the nation was grieving, and large crowds had gathered in front of the royal palace to commemorate him.

"Mom bought lots of flowers, and we traveled to Oslo from Jessheim to light candles in front of the palace," Else says. "And then I sort of realized that my mom was upset about something else. She was acting differently to the other people there. Because it was about her."

A year of hospital admissions and discharges went by, and somehow her mother was never quite the same; she became a shadow of her former self. Before that, every day with her mom had been like a surprise party; now there were secrets and shadows. "People would call and want to talk to her," Else continues, "and I'd lie and say she wasn't at home, which was scary, because she was home, and I had lied. It was a little thing that became very significant."

And then one February day in 1992, Else's kitchen was full of people, and her mom was gone. Her father had to tell his children that their mother was dead. Else ran into her parents' bedroom and buried her face in her mother's bedsheets. For more than a year after the suicide, Else carried on going into her parents' room to see if her mother was under the covers of the double bed. She had left a void that no one could fill. And Else would call out to the empty bed, saying that she wanted her mother back, that she couldn't just leave.

"Yes, it was a lonely feeling—an extremely lonely feeling," Else says today. "But I also clearly remember all the people around me who tempered that loneliness. I had friends and support available to me right away, there were pastries in the freezer, I had someone there when we heard she had died, and someone there every day afterward. Classmates visited, with drawings and kind messages. And when I returned to school, I was totally open about what had happened. It was perhaps easier because I was a child. I wasn't concerned with it being taboo.

"Dad was keen on telling the unvarnished truth. For me, it was because I didn't quite understand what it actually was, but I noticed that it made people very sad when I told them, and you get tired of having to comfort people for a whole evening. I would often lie and say it was cancer, when people asked about my family, to avoid any more questions."

If someone you are very close to takes their own life, the loneliness attached to it is especially acute, because suicide is for many still a taboo.

"The feelings of loss, sadness, and loneliness experienced after any death of a loved one are often magnified in suicide survivors by feelings of guilt, confusion, rejection, shame, anger, and the effects of stigma and trauma," writes a research group at the University of San Diego. "Furthermore, survivors of suicide loss are at higher risk of developing major depression, post-traumatic stress disorder, and suicidal behaviors, as well as a prolonged form of grief called complicated grief."

The experience for the Kåss Furuseth family was very similar.

"As my brother got older, it became clear that he had the same problems as my mom," Else says. "But when he got sick, it was different to when my mother got sick, because by then I was an adult myself. I was eighteen when he first attempted suicide, so after that we had a kind of dialogue with him about keeping him alive, which lasted quite a long time. Sometimes it went well, but I often had to be on the lookout. I felt responsible for him."

It was a course of action that involved the whole family, who made agreements with the brother to try to keep him anchored in everyday life, and for a year Else moved in and lived with him. To this day, no one in the Kåss Furuseth family turns off their phone or ignores it when it rings. On one occasion, Else was taking a shower and unable to pick up the phone when her father called, so in less than a minute he jumped in the car and raced into Oslo, forty kilometers away, to check that everything was alright. Else, her sister Cecilie, and their father are a close-knit family who look after each other—and they also bond through a shared sense of humor.

Research shows that what Else's family experienced is not as extreme as you might think: the suicide of a close family member can trigger another in the same family. A large Danish study involving 1.5 million people and spanning almost twenty years showed that the risk of self-harm, mental illness, and suicide increased after the suicide of a loved one, especially in the first year after the loss. An English study involving 429 survivors of suicide tried to establish what the trigger for this is. Most of the bereaved described themselves as having an awareness of suicide and were very clear that they didn't consider it an option for themselves; but for a minority, suicide became a possibility they hadn't considered before.

Both the suicide and the prolonged grief are related to loneliness; both create a strong feeling of not being understood, that you are totally alone in all that darkness. This was something Else wanted to investigate. So she made a TV series about suicide, talking to people who have attempted suicide themselves.

"People often say it was a little thing that made the difference. That someone from the football team rang, that they felt like someone cared," she says. "But I don't know if that's always true.

I don't know if there was anything that would have prevented my mother's and brother's suicides. I do believe in a way that his suicide was a consequence of hers, that it was the trigger, not that she was to blame in any way. But I'll never get an answer to that. And I don't really think they could be saved. At the same time, I do, of course, think I could have prevented it; we feel guilty because we take responsibility for each other, that's how it is. Everyone has responsibility for their own lives, but I think we underestimate how important we are to each other."

In her book *Else går til psykolog* [Else goes to a psychologist], she describes her attempt to process the complicated grief that the two suicides left her with. Complicated grief does not follow a "normal" grief pattern, from open wound to slow healing. Instead, the open wounds of anger and confusion do not heal; they remain there, actively hurting. A suicide also feels like a form of rejection. It is loneliness, both for the person who ends their own life and for all those who loved that person and tried to keep them alive.

"Nothing will be the same again. It's stupid to leave, but they left me. I am not enough," Else writes, describing the strong feeling of abandonment a bereaved person can feel. It has made her a strong advocate for openness and community, and she has in the past commemorated World Mental Health Day and World Suicide Prevention Day by opening her home and inviting people in.

"Hand on heart, I believe there would be fewer lonely people If there were more people living together in small areas," she says. "I genuinely believe that community makes people feel better."

People who attempt suicide are usually characterized by feelings of hopelessness, desperation, and isolation from family and friends. Depression and bipolarity, as well as drug addiction and schizophrenia, are among the most common underlying mental illnesses, but trauma, rejection, loss, and disappointment can also trigger a suicide.

"We need to remember that suicide is nevertheless rare. We humans are equipped with numerous barriers to it, even though most of us can have suicidal thoughts and are aware of the existential questions related to it," says Fredrik A. Walby, a specialist in psychology who researches suicide.

But one thing is clear: if you are closely involved with your community, the risk drops considerably. "The better integrated you are into society, the less risk you have of being suicidal: if you are married, have children, and have a job, you are definitely in a low-risk group," Walby points out.

This doesn't mean that statistics can explain everything, or that being seemingly connected protects you from suicidal ideation. Walby's research shows that there are clear risk factors as well as a lot of mystery connected with suicide. Risks occur during major upheavals in life, during a breakup or after the loss of a job: all of life's dramatic transitions are connected to potential loneliness. Severe depression is a risk. Not talking to anyone about how you feel is a risk. Addiction is a risk. Feeling ostracized and lonely is a risk as well: the loneliness can become so great, and the sense of not belonging in the community so extensive, that you can't take it anymore. At least, that's what the researchers think. The problem is that people who attempt suicide but don't succeed aren't necessarily the same type of people as those who do succeed. At the same time, research shows that many of those who do succeed have tried once already. The risk level of suicide is related to the number of previous suicide attempts.

But major societal changes, such as a pandemic, can also affect the suicide rate—though not necessarily in the way you might think. At the end of the nineteenth century, the renowned sociologist Émile Durkheim launched a thesis about how the suicide rate would decrease during wartime. One of Durkheim's seminal works, *Suicide*, from 1897, would have its hypothesis tested only seventeen years later, at the outbreak of World War I. David Lester at Stockton University set out to investigate Durkheim's thesis and found that Durkheim was actually right: during both world wars, the suicide rate went down a little, not up, as one might have assumed. For both men and women, on both sides of the front line, the suicide rate fell—even in countries that didn't take part in the wars.

When faced by an external threat, we bond more as a community, we unite against an external foe. And this can be an antidote to loneliness—an anti-suicide medicine. "I see no reason for a pandemic to make the suicide rate go up in the short term; it should actually decrease," says Fredrik A. Walby. "For those living with a lot of inner turmoil, it's in a strange way reassuring when a crisis hits, because they are no longer experiencing it alone. They get to be part of a collective and rational worry instead of brooding over their own familiar traumas."

When the crisis hits the community, the person suffering from depression and anxiety is finally understood and able to understand their surroundings. It creates a long-awaited sense of community. But it was not like that for Ada. When crisis hit Oslo, the darkness she was experiencing simply worsened.

Let's return to the bomb.

What I didn't know when it exploded was that the man who had left the huge bomb outside the government headquarters before strolling calmly in a police uniform toward the getaway car—the man who would later shoot Adrian Pracon in the shoulder—had been in my life before, that very same summer. I had met him.

It was only after all the details about his life and activities before the terrorist attack were published that I realized that in the summer of 2011 we had both been regulars at the same nightclub: Palace Grill. There had to be a reason why I remembered the episode, something about him must have stuck in my memory, because I had met plenty of other people that summer. But one news article reported that he would say the same thing to everyone he met that summer—and that's why I think it must have been him.

Here's how my conversation with the future terrorist went:

It was a warm summer evening, and we were standing in Palace Grill's backyard, surrounded by brightly colored string lights. I was often there that summer and had met and spoken to a lot of people, but I remember him: tall, ordinary, slightly measured, and typically Nordic, like so many other men out on a Saturday night. We stood there watching the crowd of people dancing. Like me, he clearly wasn't interested in throwing himself onto the dance floor and, like me, didn't seem to have an alternative plan. So I looked at him and said, "Nah, no dancing for me," and he agreed—and then we stood for a while in silence, observing the party.

I asked him what he did for a living. I'm sure I smiled at him, though I don't think he smiled back, which wasn't that strange, given that Palace Grill was a pick-up joint for hipsters and people with money. It wasn't somewhere you wanted to appear too inviting and open.

"I'm writing a book," he said.

"Oh, interesting," I said. I was also writing a book that summer, my novel about longing and loneliness. "I'm writing a book too! What's yours about?"

He was evasive and didn't answer.

"Are you getting funding for it?" I asked, thinking I might be able to help. "I know a lot about the kind of grants you can apply for!"

I didn't know then that the manifesto he was writing was mainly cut and pasted from other people's conspiracy theories and garbage on right-wing-extremist websites, along with snippets of his own quasi-heroic life story: 1,518 pages of hatred and loneliness, which were to be distributed by email once his car bomb was complete and ready to go. It wasn't exactly the kind of material that would get support from the Norwegian grant system.

"I don't need a grant," he said proudly and dismissively, "I've got plenty of money."

And that might be what I remember finding strange: a writer who wouldn't talk about his book with another writer and who bragged about all the money he had. That doesn't happen very often. He was strange, but it was hard to detect behind the slick, west Oslo facade. He seemed so normal on the surface. But he also had an aura of coldness about him, or maybe not even that, just ... nothingness. That's what was so strange, and maybe that's why I remember him: it was like he didn't give off any scent.

Anders Behring Breivik grew up on Oslo's prosperous west side, but his family was anything but affluent. When he was little, his parents separated, and the years that followed were likely tainted by violence and abuse, with little Anders trapped in the flat with his mother. Her own childhood had been rife with serious neglect, a dark legacy she passed on to her child. She had Anders in 1979, at the age of twenty-six, but quickly lost contact with his father, Jens Breivik, a diplomat.

In 1981, when Anders was barely two years old, his mother contacted child services for the first time: she needed help with Anders and his sister. Then in 1983, she contacted child services again. She believed that Anders was "hyperactive, clingy, and passive." The psychologist who examined him wrote that he had become "a somewhat anxious, passive child who averts contact, displays a manic defense mechanism of restless activity, and has a feigned, deflecting smile."

This is a description we easily recognize from the Strange Situation as disorganized attachment. The restlessness is connected to hypervigilance—the fear of danger and overactivation resulting from an unpredictable caregiver. It is quite possible that the anxious son had to deal with an adult who was loving one moment, dismissive the next. In the description of Breivik's childhood, there are hints that there may have been sexual abuse by his mother, but no proof, of course. It is even reported that his mother referred to him as "a difficult child" when she was carrying him in the womb. The psychologist recommended that this child—whom his mother considered a problem even before he was born—be taken out of his family. But it didn't happen. Even when Breivik's father sued for custody, the four-year-old boy wasn't taken out of the home. It was also reported that the boy and his sister were regularly alone in the apartment while their mother was out, even at night. So we can only assume that this child remained in a situation rife with psychological violence and abuse, without anyone intervening.

"He tried to form an identity by shopping for brands in the stores and for ideology online," writes Aage Borchgrevink in the book *A Norwegian Tragedy*, about Breivik and his radicalization process. "The attachment disorder evolved into what the medical examiner's office described as a complex personality disorder with an emphasis on narcissistic, paranoid and dyssocial features. In the manifesto, this comes out clearly, together with his misogyny and sadism."

Borchgrevink points out that "the psychiatrists who assessed the family described the boy as having what is now called an attachment disorder. He had normal abilities but was unable to imagine other people's feelings or understand himself. The psychiatrists' description is similar to how Breivik in the manifesto depicts himself as a zombie, living dead and without ties to other people."

Borchgrevink has carefully read the manifesto, which Breivik emailed to 1,003 recipients before driving the white transit van into Oslo. To Borchgrevink, it seems that Breivik's narcissistic traits made him incapable of separating his own inner universe from the external one, as though he projected his own traumas onto everything around him. The entire world was his mirror: "It is typical of Breivik how he projected his individual experience to apply to the whole world. His mother was 'emotionally unstable,' so it followed that all women, or at least 90 percent of them... were 'emotionally unstable,' and did 'not understand codes of honor,'" Borchgrevink writes. "He was unable to distinguish between his own inner world and the external one, between his own, possibly forgotten, trauma, and the varied and nuanced landscape that is society. Everything was equal, and nothing existed outside his own mind."

Borchgrevink believes that the atrocity on Utøya and in the Government Quarter could have been avoided had Breivik been taken from his mother when he was small, as the experts had recommended. "It doesn't mean that the terrorist attacks weren't political, or that we don't all have a responsibility as citizens to react to extreme expressions and actions in our own society," he clarifies.

Anders Behring Breivik never became part of a community. He tried to become a tagger but wasn't accepted by the other taggers. He joined the youth wing of Norway's Progress Party, but failed to make any impression and quickly gave up party politics. He sought out like-minded people on far-right websites but was overlooked there too. He also spent a lot of time playing *World of Warcraft*, claiming it was part of a "desensitization process" before the murders. But it's possible that he didn't need to be desensitized. He may have been completely numb already.

When he went looking for someone to direct his frustration and hatred at, he found Muslims and women; all those who stood for diversity and an open, nuanced, and multicultural society; all those who wanted a world that isn't black and white and easy to handle. He chose right-wing extremism, which offered a system tailored to his own paranoia and distrust of the world. Research on right-wing extremists shows that many of them experienced violence as children.

Symbolically, the main targets of the terrorist attack were two women: either the former prime minister Gro Harlem Brundtland (known as "the mother of the nation") or the journalist Marte Michelet were to be ritually beheaded, and the film clip posted online. Much of Breivik's manifesto revolved around gender and gender roles.

When I met Anders Behring Breivik that summer night in 2011, I didn't know who he was. I didn't know that I had met someone who was as lonely as me. That he was my mirror image. He could have been a character from Moominvalley, the Groke even. I remember how the conversation simply died because there was nothing about him for me to latch onto, and then the soon-to-be-world-famous man got up from the bar and walked to the door. Could I have stopped him? As I remember, he became a shadow who just slipped away.

Many years later—among the chaos of dismembered dolls and bedraggled teddy bears in my daughter's toy box—I found some round pieces of wood with holes in them. They were ochre, grass green, red, and light blue, and the paint was cracked and peeling. Intrigued, I scooped them all up and found the narrow stick they were supposed to be threaded onto. It turned out to be a little wooden tower, perfect for children wanting to practice their motor skills—and also a small and helpless symbol of how community is greater than its individual parts. My daughter, who was seven at the time, was clearly done playing with it and had just tossed it unceremoniously into the toy box without any more thought.

I don't know the exact origins of this wooden toy, but it was likely made by my grandfather in the 1940s, when his children were young, though it's possible that he made it earlier. There's no one I can ask about it now, but I seem to remember hearing that he learned how to operate a lathe during his time in a labor camp in Norway's far north. He had been a prisoner of conscience during the Nazi occupation and considered a sort of hero in my family. There was something glorious about his life, a destiny that began in humble circumstances and went on to become part of the struggle against fascism. But let's start at the beginning, let's start with the whale hunter.

My great-grandfather was born in 1902, into a century marked by great conquests and the trauma of war. Over the course of a long life, he saw the ideas of the nineteenth century tested and stretched to the breaking point by the crazy events of the twentieth century, while alienating but marvelous cities and mass production expanded at the same rate, consumption increased, and amazing inventions made life easier and more mobile. Women and workers gained rights; prosperity increased for everyone. But he also witnessed the enormous Spanish flu pandemic that killed fifty million people, and two world wars that left another ninety million dead and even more wounded and traumatized. If anything, the twentieth century was the century of extreme violence and sudden liberation.

I suppose that's how it was for my great-grandfather. The possibilities back then were huge: he could become something entirely different from a farmer's boy, he wasn't bound to become one. He came from a small farm and his family was poor, so he was sent temporarily to another farm. This was quite common at the time, but it would have no doubt affected him and given him a sense of abandonment. When he was fifteen, he signed up to work on a whaling ship bound for the Southern Ocean, where the large pods of blue and humpback whales kept the processing ships busy day and night. Did he see a lone whale out there among the icebergs? Was there a whale singing at 52 hertz back then, too, one that couldn't talk to the other whales? This man would one day give me something, I think: perhaps my curiosity, my restlessness. Perhaps my ability to feel lonely.

Searching for the origin of your own loneliness is like scouring the Southern Ocean for a rare whale—a 52, a Moby Dick. I'm in cold, foreign waters and it's hard to find any tangible clues to follow. I want to outline the kind of loneliness that is making me sick and stressed, that has given my life a dark undertone. But it's hard to pin this onto any map. I have learned, while working on this book, that attachment patterns are inherited. So what have I inherited? When my great-grandfather was a teenager, he left the farm he'd been sent to and went to work on a whaling ship. What did that do to him? Life on a whaling ship consisted of grown men, a rough environment; it was no place for care and emotional growth, no place for developing the capacity for love. Masculine cohesion was paramount. Did this experience cause a wave that is still in motion today—moving through me? Is my great-grandfather's boat the stone in the water?

His son became a teacher—was *he* the stone in the water? That can't be right, because my grandfather was a hero. When the occupying power seized control of Norway in 1940, its attention soon turned to what all propaganda machines want control over most: schools. We see this in all totalitarian regimes. Winning the children is winning the future; dictators want to control dissent and criticism before it can be formulated. The highest Nazi authority in Norway, Josef Terboven, demanded teachers sign a declaration of loyalty to the new regime. But the teachers refused to sign. So, at the end of March 1942, my grandfather was one of 1,100 male teachers who were arrested, of which 641 were put into cattle wagons, 45 men in each. By the time the wagonloads of teachers were about to leave Oslo, the rumors about who was inside them had already spread. And this attracted children from all over the city, who gathered around the train cheering and singing the national anthem. It was a display of support that continued along the entire stretch between Oslo and Eidsvoll, a hundred-kilometer line of cheering and singing children.

The other prisoners were treated far worse, of course. While force was being used against Norway's primary school teachers, the Nazis were launching the "final solution" that had been planned at the Wannsee Conference on January 20, 1942. The ongoing genocide of the Jews had from then on become more organized and targeted than before, driven by racism, antisemitism, and a conspiracy theory based on the contents of the fabricated book *Protocols of the Elders of Zion*, from 1903. But the Nazis didn't want Norwegian Aryan teachers on their conscience, so my grandfather was in the end given relatively civil treatment. But he wouldn't have known that as he approached the prison camp in which he spent the following months. He just knew that he was a prisoner and couldn't have known when he might be free, or if he would ever be free. I don't know if my grandfather got PTSD and trauma from being locked up, but he came home with no physical injuries. I don't know if he went on with his life unaffected, teaching in a small village school, or if his experience remained hidden beneath the surface; he never mentioned any of it.

So many secrets, so much hidden loneliness. Violence is like a stone dropped in a lake, triggering circular waves. Maybe I'm at the outer limit of one such wave? What if my daughter is in turn affected by me, what if she notices the vibrations? How could she *not* notice? These waves are over a century old, and have survived two world wars, so it's hard to be a breakwater. But I want to be a breakwater.

American psychiatry professor Rachel Yehuda has researched how trauma can move from one generation to the next, before a child is even born. She studied seventeen hundred women who were pregnant during the September 11 attacks and found that those who suffered PTSD afterward passed their high cortisol levels on to their children. But trauma is mostly transmitted in completely different ways than with cortisol. One of the most important reasons why children are exposed to violence is that their parents were exposed to violence. It is a vicious spiral of violence carried out by victims of violence, thus creating even more victims of violence. This perhaps isn't so surprising, since we know that the coping strategies children learn at a very early age can stay with them their entire lives. Being exposed to violence as a child also makes people worse at handling all types of emotions through "mentalization," the ability to understand others. And it's also possible that using violence transfers pain. In an experiment with rats, which subjected the poor animals to electric shocks, what eased the rat's pain slightly was allowing it to bite a stick. However, what reduced the animal's pain even more was putting another rat in the cage, which the tormented rat could itself torment and bite. Transferring the pain and humiliation can perhaps make the violence easier to bear. Similarly, excluding another person can offset your own pain of being ostracized.

I remember how, when I was my daughter's age, I would hit other children. I remember my clenched fist striking the child's back, the sound of a hard fist against a hollow rib cage. The sobbing child. I remember the relief it gave me. It's something I still feel ashamed of and something I've never mentioned to anyone: I was transferring the abuse, and it made me feel better. But it also made me feel lonely—everything about that situation was lonely. I shoved people away who could have been my friends, while knowing that I didn't belong in any community. Violence is fundamentally lonely.

Being exposed to violence as a child also makes us worse at handling all kinds of emotions and less able to connect with other people. Family events a century old can shape the relationships we have today. Perhaps the terrorist Anders Behring Breivik was living with trauma that began several generations earlier. I'm starting to realize that what I've experienced myself is an echo of someone else's pain and anxiety.

"I have that loneliness within me. It's something I carry and always want to keep," says Madeleine Schultz, who has written the book *Morslinjer* [Maternal Lines], about how violence repeats itself, one generation after the next. Schultz's grandmother chose a man who treated her badly, and her mother was abused as a child, which in turn has affected her ability to touch and comfort. When she discovered that this is a pattern that repeats itself, she had to write about it. "It helps to accept that it's there, and that I'm not the only person feeling this way," she says.

When Schultz's grandmother divorced her abusive husband, she was left poor and very alone. And throughout Schultz's childhood, her grandmother constantly repeated: "Nobody realizes how much pain I'm in." It was a mantra she repeated so often that in the end nobody listened to what she was saying, it just became a noise in the background. But after her death, Schultz found all the letters her grandma had written to the Norwegian authorities, desperate letters about being unable to feed her own children. She was also very sick. But there was never room for her suffering. She was never listened to. She was one of many nameless and unimportant and impoverished women, whom no one tried to help.

"She was so lonely, and she couldn't put it into words," says Schultz. "She became this bitter old lady, she didn't like anything, didn't want to travel, she just totally closed herself off, it was a type of loneliness that was all-consuming." Her grandmother's family had moved from the rural north to Norway's capital and ended up distinctly working class, living in such poverty they had no chance of understanding and mastering the city's social structures—there was no way for her grandmother to break free of her economic troubles. Once the drunk and violent husband was gone, she became financially helpless.

"There is so much ignorance about violence: people think we're either kind or mean, but people can be both kind and violent. People can be more than one thing. Failing to understand that makes us blind to the violence. And this blindness means that the victims' stories usually remain untold," Schultz says.

"To overlook these stories is to overlook the fact that it is our mothers who teach us what love means, or conversely: what it means to be always longing for it," she writes in the book. "Shame and loneliness lock the violence in, make it invisible."

Schultz is the third generation. The previous two generations of women were trapped in economic unfreedom, browbeaten and abused by men. But Schultz now thinks she has broken the pattern. She has studied art, she has worked as a journalist, and she has worked hard on confronting her own traumas. Today, she has a husband, a five-year-old boy, and a baby girl.

"I've thought a lot about how my children must never feel the same loneliness that has followed my family for generations," she says. "It's something I was very worried about when I was pregnant. I might look at my son from a distance sometimes and notice an air of sadness about him—but it's dangerous to project your own feelings onto a child. I know that he has two very caring parents, and I've worked very hard on myself: if you're aware of these structures, you've already done an important part of the job."

Schultz's experiences corroborate something that became clear from attachment research in the 1970s: that a child's inner world can be full of ghosts and dark shadows from the past, and that if you understand where these shadows are coming from, they become less dangerous. A large percentage of children who are exposed to violence become violent parents. Stopping violence in close relationships might stop it for generations to come. And one of the most effective ways of stopping it is *self-reflection*. Professor Selma Fraiberg and her research team at the University of Michigan investigated how a bad childhood could be "handed down." How was it that an adult who was neglected as a child could end up giving their own child a similarly insecure upbringing?

"In every nursery there are ghosts. They are the visitors from the unremembered past of the parents, the uninvited guests at the christening," she writes in the introduction to the ground-breaking article "Ghosts in the Nursery," from 1975.

"Even among families where the love bonds are stable and strong," she continues, "the intruders from the parental past may break through the magic circle in an unguarded moment, and a parent and his child may find themselves reenacting a moment or a scene from another time with another set of characters. Such events are unremarkable in the family theater, and neither the child nor his parents nor their bond is necessarily imperiled by a brief intrusion."

The problem, of course, is when children are constantly visited by their parents' own ghosts from the past, those which are driven chiefly by the parents' unprocessed memories: a mother who doesn't relate to the violence she was exposed to herself will be unable to see the vulnerability either in herself as a child or in her own child. In the article, Professor Fraiberg follows a young mother with trauma who, despite her good intentions, has been pushing her child away and been unable to give it the closeness it needs. The research team then helps the mother to confront what has happened to her. Their theory is that becoming a parent is, in a way, about embracing both your own inner child and the child in front of you. Only then can there be pure love between parent and child, a totally unconditional love with an open heart.

Fraiberg didn't yet have the tools for testing her hypothesis, but in 1993 a thorough empirical investigation of her ghost theory was carried out. In the article "Measuring the Ghost in the Nursery," Peter Fonagy, a clinical psychologist at University College London, closely examined both children and their parents. The children's attachment patterns were measured using the Strange Situation test, and the adults were evaluated using what's called the Adult Attachment Interview (AAI), a comprehensive interview where both the answers and *the way they are given* are important to the researchers. The interviews were conducted while the parents were still expecting their first child, so that an evaluation of the parents' attachment style had been made before the examinations of the children began at twelve and eighteen months. It turned out that what parents revealed about their own attachment style enabled the researchers to predict what kind of attachment style the children would get: it was something the parent handed down.

An especially important point for the researchers was how coherently the adults talked about their own childhood experiences. It seemed that having a clear and comprehensive story about what happened to them as children, whether it included bad or good experiences, affected how secure these adults felt when they became parents. For example, a mother-to-be could say that "security was very important in my family" and in the next breath say that she used to be beaten, thus presenting a paradox. It meant that she'd been unable to create a comprehensive and coherent narrative about the violence she had experienced, because violence is not security.

"We found that, in the parents' responses to the AAI, the factor which best predicted insecurity in the child was a parent's incoherent narrative about their own childhood," writes psychologist Fredrik Cappelen in an article citing Fonagy's study. "This is in line with Fraiberg's hypothesis in 'Ghosts in the Nursery'; parents' unprocessed experiences, or unintegrated experiences, experiences characterized by defense and repression, have the most significant effect on their attachment to the child."

For the mothers who had a secure attachment style in the Fonagy study, 78 percent of the children had a secure attachment style. For the mothers with an insecure attachment style, 72 percent of the children had an insecure attachment style. When it came to fathers, it turned out that secure fathers produced 82 percent secure children. Insecure fathers were somewhat less likely to have insecure children—only 50 percent.

A child who feels insecure in their relationship with their parent turns away from the parent and finds it more difficult to regulate emotions and stress, and also becomes a stranger to themself. The child doesn't really know what they feel and need. They become lonely.

"They also imagine that the experiences create a sense of estrangement from the self, something they call experiencing the *alien self*," writes Cappelen. "The *alien self* is often noticed as feelings and experiences that you don't understand, such as self-estrangement in the form of self-harm, outbursts or self-criticism."

Attachment psychologist Ida Brandtzæg adds that therapy and help with interacting with your child is available and has produced very good results. "We tend to pass our own childhood traumas on to our children," she says. "But there is a way out of it. You don't *have* to repeat your parents' mistakes. It's quite clear that reflection on your own relationship experiences and good social support can enable you to break out of it."

Brandtzæg and fellow attachment psychologist Stig Torsteinson aren't just close collaborators; they are also a couple, with six children from previous marriages to look after.

"With all you know about proper attachment, do you manage to do everything right?" I ask. "Aren't you afraid of doing something wrong? What might I be doing wrong?"

"No. It's not about being a perfect parent, just *good enough*. We make a lot of so-called mistakes," they both say reassuringly. "All relationships, even secure ones, have repeated breaks in the interaction. But in secure relationships these breaks are repaired more often. Things get back on track."

We so easily think that it's an either-or answer, either a very good upbringing or a very bad one, but it seems that the right answer is somewhere in between. The children who show the best attachment are those who have been moderately understood by their parents, not those who are given constant attention and understanding. How sufficiently a parent connects to a child's needs can have a future effect on the child's ability to regulate stress, how it develops empathy, how secure it feels in the world, and how confident it will feel in being cared for and understood. We can all comfort Toffle, it's not an impossible task. But when children and teenagers go for too long without feeling loved, they start to feel emotionally cold. They become the Groke. Our goal must be to have as few Grokes, as few invisible children, in the world as possible.

Ada and I understood each other's sense of homelessness and being unwanted, and this loneliness united us. But now she is gone, and I'm one of the few people who remember her; and the only evidence of me knowing her is a long black strand of hair wedged between two floorboards in a flat I lived in ten years ago. Why am I alive and not her?

Of those who experience such darkness, most choose life. And there are many of us. The vast majority of lonely people are like me. We claw our way back into society and give life another try. We stay. We don't make bombs or plan terrorist attacks, we don't jump, we don't give up. We instead carry the grief and the violence and the humiliation within us, as dark impressions and inexplicable pains that cannot be cured, like insomnia and anxiety, poverty, exclusion, and addiction. We reach out to other people, even if it hurts. We don't commit acts of violence. Studies by the University of Oslo's Center for Research on Extremism (C-REX) show that even right-wing extremists hold each other back; they stop each other from committing acts of violence or terrorism.

I know who Anders Behring Breivik is, because he is my mirror image: while he was writing his pathetic manifesto, I was writing a book about humankind's longing for love. In the original draft, my story ends with the protagonist blowing up Oslo's literary arts center. But after the July 22 attacks, I realized that I had to write a different ending, one that wasn't so frighteningly similar. In the new draft, my protagonist walks into the sea to drown herself, then turns and walks back toward the living, embarrassed but alive. There are several ways of dealing with emotional pain. We can pass it on as violence and terror. Turning the violence on other people might bring you temporary relief—*finally*, you think, finally others can hear your scream of pain. Was this what Anders Behring Breivik thought when he lit the touch paper and calmly walked away from his car bomb? Did he afterward think, "Can you hear me now?"

We can also turn the pain on ourselves, until it becomes unbearable, as it did for Ada.

Or we can carry our pain through life and endure it, with all the loneliness and strain on the body that entails, like a constant pressure in the chest that you must learn to live with. I *must* learn to live with it, despite the pain; I have to be a good enough mother.

When my father died, I lost someone who actually did try to keep me warm during my childhood. He was my security blanket, the person who loved me unconditionally. Now he is gone forever. But every day, the little warmth I got from him emboldens me to try to break the vicious circle. I still find it painful to think about. Because when he died, he went through the most profound loneliness a human can experience. He went all alone into the unknown. I hope my singing softly at his bedside helped. But I can't know, because he never woke up. At the same time, I suddenly realized how vulnerable this idea of the totally independent and self-standing individual actually is. This idea, which began in the Renaissance as an escape from the collectivist Middle Ages, has now become almost like a straitjacket: in the secular world, it makes death so much harder to contemplate. It means single-handedly carrying your entire world, and your death is its complete destruction. But if we are each an irreplaceable part of a strong community, death need not be as devastating: the community will continue our legacy.

So my father isn't totally gone, because many of us will carry pieces of the good he was and did within us into the future.

While working on this book, I have wondered how to convey my own loneliness. I struggle to understand why anyone would want to read about it, because I find it so uncomfortable and difficult to write about. Each step I've taken into the land of loneliness has felt like dragging my feet through a quagmire. Loneliness doesn't show up on my map, it isn't documented; it is hidden in the corner of my eye, in the corners of my life; it is a map full of white spots. Loneliness is still uncharted territory and hard to get an overview of, and when I try to draw it on a map, it spontaneously combusts.

I found that it's impossible to tell a story about loneliness without telling many other people's stories; our lives are so intertwined. The stories about me are about everyone I know as well. They are the stories about those who failed and forsook me and told me I didn't deserve to be loved and protected, but they are also the stories of the many, many people who held me tightly when I needed it. All my experiences of loneliness have changed me and still affect me; loneliness is connected to the darkest points of my life. I have known that I don't belong, that I am lost, that I'm not needed anywhere, that I'm a leftover. Loneliness is my permanent setting. My husband was right, of course, I really am the loneliest person he knows. Loneliness is waves of dark water, radiating out to me from generations ago. I can only pity the ancestors who handed the violence and coldness down—an entire family of Grokes.

I think of all the places where loneliness hides. Another form of loneliness is, of course, heartbreak, because it comes from losing someone you thought would be your soulmate, who was supposed to be right next to you. Any pop song or movie or poem about heartbreak is about loneliness too. Almost every story in the world is about losing community and finding community, about abandonment and longing and safe harbors. I suddenly understand how loneliness and religion are connected: God understands and protects you, promises to be with you when everyone forsakes you, protects you from loneliness. But loneliness disguises itself as envy and jealousy; it hides in pointless consumption and overeating, in shame and bullying and sarcasm, in the pursuit of status, in violence and abuse and racism, in never-ending clicks on social media. Loneliness exists where people are poor and cannot help themselves, among the sick and dying. It's the absence of touch or a hug; it is sitting alone in a room in front of a screen. For a while, I thought about writing a list, a summary perhaps, a cartographic record of the solitudes in my life, I felt that I really ought to mention everything. But then I realized that I didn't need to, because all the stories in this book are stories about my loneliness too. In their own way, they mirror mine. In their own way, they mirror yours.

After realizing that I'm a Groke, every day has become an exercise in asking for help, in not going on the offensive when I feel clumsy and helpless, in allowing myself to be vulnerable, in asking for a hug. I will ask for a hug and stay in my husband's arms until the fear lets go of me. When an argument is brewing, I'll say "Sorry" and "It was a misunderstanding," instead of going on the defensive. I try not to work all the time, I take weekends off, which at first makes me anxious—I feel useless and unworthy—but eventually I start to relax and connect with the people around me. I laugh with my daughter so hard she cries out, "I'm peeing myself!" I smile at my husband for no precise reason. It is unusual. It is nice. It is scary.

After all the research I've done, I'm not surprised that we felt gnawed at by loneliness and depression during the pandemic, because withdrawal is withdrawal, whether it's chosen or not: whether we have secure attachments or not, the pandemic forced us into a temporary period of almost mechanical depression and loneliness. It forced us apart, in some cases for years. Afterward, those of us who had the network and strength to do so could finally brush off the pandemic and move on. But those of us who don't have extra resources might just let go and let ourselves sink to the bottom.

We all have different capacities for feeling lonely; some of us have previous trauma, some of us are naturally avoidant and introverted. But the statistical patterns are clear to me; there are some paths that loneliness does follow. It's no wonder, for example, that people living in institutions—such as care homes, hospitals, prisons, and psychiatric wards—feel lonely, because to be institutionalized is to be taken from all you once knew and put somewhere totally alien to you, beyond society as a whole and any social circle. It feels like a threat of ostracism. In prison, you will likely also experience a degree of trauma and violence behind bars. And if you are overweight or have a skin color that isn't white, if you have a mental or physical disability, if you are poor, unemployed, or on social security, if you have a sexual or gender identity that differs from the majority, there is a high likelihood you will feel lonely. And this loneliness isn't something that just suddenly happens. Ostracism follows a pattern: the so-called weak, those whose bodies and lives deviate from the norm, those who are exposed to violence and are victims of bullying feel acute loneliness because *they really are ostracized. They become lonely.* Society exists for the strongest. Society exists for the well-adjusted and wealthy, the hardworking; and here in the West, it is those with skinny white bodies who effortlessly rise up the social ladder. Society, which was designed to look after the weakest, hides brutality and loneliness. The vague sense of exclusion spreads, from glances and laughter and body language, from not being understood and heard, to not being held and cared for by institutions or fellow human beings. And then, as a logical consequence of all the prior exclusions and signs of how unimportant and shameful you are, come the violence, the rapes, the murders, and the suicides. In that first contemptuous look, there is the threat of violence.

I believe that many lonely people respond adequately to countless small and large rejections, but so begins the withdrawal from the community, and the maelstrom swirls into action. And this sense of uncertainty is also the driving force behind so much of the consumer pressure we feel: there's always someone who will capitalize on our addictions and insecurity. But while capitalism is part of the problem, it is only part of the problem. Capitalism exploits the innate brutality of humankind, our desire to rise in the hierarchies by having a prettier face or a smarter head than those around us. And what we don't already have, we buy. When the pain is unbearable, we happily pay for medication in the form of alcohol, drugs, food, clothes, and plastic surgery. We buy cars, holiday trips, and other status symbols to show how independent and self-sufficient we are. The independent, self-sufficient man is a strong individual, a non-lonely man, someone who only needs other people for admiration and validation.

The author Haruki Murakami wrote: "Why do people have to be this lonely? What's the point of it all? Millions of people in this world, all of them yearning, looking to others to satisfy them, yet isolating themselves. Why? Was the earth put here just to nourish human loneliness?" No, the planet doesn't nourish human loneliness; we have hands and hearts that can undo it and break the dark spirals. We can all be warm woolen blankets in each other's lives. We can be like the inhabitants of Moominvalley and sit around the kitchen table in the Moomin house. We can become breakwaters. We can be the Groke dancing on the sand.

Many paths can lead to loneliness, it is a maelstrom you can be sucked into in different ways: a subtle rejection at school that can make you retreat to the bedroom and start gaming. Profound sadness can make you withdraw from those around you and in turn make them withdraw, thus creating an even bigger vicious circle. You may be an introvert and unable to meet society's demand of being exciting and outgoing. You may have experienced bullying, violence, and abuse and become afraid of the world.

Instead of being surprised by how many lonely people there are, we should view them as symptoms of a society where exclusion and violence is accepted, where those who are silent about their needs go ignored. We live in a society where one person's pain is worth more than another's. So, to me, it seems like society needs to change more than the people who are lonely. For me, it's a wonder there aren't more lonely people in the world, when so many people experience violence and abuse and racism and bullying, and so many live with trauma and grief, so many live without hugs and eye contact, so many people are poor and afraid, invisible and hidden! But most of us carry on nevertheless. We don't give up. We are drawn to community by forces more powerful than those that draw us away from it.

July 22 was a gray and cold summer day. And the fertilizer bomb that Anders Behring Breivik had spent months making was ready. Everything had been planned in detail, and none of these plans had been shared with anyone—not one clue anywhere that might reveal what was about to happen. One man can be like a locked vault when he doesn't share his secrets.

As Breivik parked his car bomb outside the seventeen-story tower block, Ada was passing by on her way home. Yohan Shanmugaratnam was also there, wheeling his two children down a little side street just meters away, after visiting his wife's office to print some tickets to a Captain Hook theater show. Martin Eia-Revheim, already rushing from Grønland to a delayed meeting in Akersgata, picked up the pace as he approached the Government Quarter. Hilde Susan Jægtnes was boarding a train to Berlin. Ika Kaminka was drawing breath to duck beneath the water at the local swimming baths. Peder Kjøs was on his way downtown with his wife and kids to watch the *Bob the Builder* movie. Adrian Pracon was at Utøya, where the politician and public speaker Ali Esbati was giving a talk about right-wing extremism. Lasse Josephsen was hunched over his computer in Arendal. Shazia Majid was on her way to a dinner with friends. Helene Flood Aakvaag was buying a coffee at Oslo Central Station and would soon be driving past Utøya on the bus. Marthe Bødtker was driving into Oslo with her son, rounding the hillside bend at Ekeberg that gives you a view across the entire city.

All over Norway, people were doing the very ordinary things that hold society together, such as serving food, or emptying the trash, or drinking beer with colleagues, or standing on a mountaintop with a group of friends, or driving the 37 bus through town, or packing for a holiday, or filing a document at a government ministry, or preparing for a Captain Hook theater show, or gathering for a lecture on racism and right-wing extremism. Beneath the water at the swimming pool, Ika Kaminka heard a loud bang. Marthe Bødtker saw a large cloud of smoke suddenly rising from the city center. Martin Eia-Revheim took off his headphones and stood amid a sea of glass that had been blown across the street. Yohan Shanmugaratnam could hear nothing but a loud ringing in his ears as his children squirmed uneasily in the stroller.

Our lives are so intertwined that we don't even notice until something huge and terrible tears us apart. Until a deep crater has been blown in our day-to-day lives, and glass is cascading from above.

One sunny day, my daughter and I go for a swim at the local outdoor pool. We ride the 37 bus across town, passing the bomb site, still off-limits and rumbling with bulldozers. It is the city's biggest open sore, a gigantic hole and daily reminder of the hatred of one lonely man, who tore our city and our country apart. Even now, twelve years later, the devastation hasn't been fully repaired. The vibrations of the bomb will stretch into the future, for several generations perhaps. My daughter wasn't yet around when it happened, not even in my dreams. But now she is climbing the diving tower and about to jump from three meters up. She runs to the end of the diving board, right up to the very edge. Then she turns and walks back, looking slightly embarrassed and crestfallen. She repeats this ritual a few times, then runs determinedly toward the edge of the board. She pauses there, one second too long, before falling reluctantly into the deep turquoise water.

I think of Ada, of how she must have hesitated for a second before saying farewell to humanity, giving up on the idea that someone could help her. She must have hesitated, just a little, before letting go. Why did she no longer believe that anyone could help her?

I think of Anders Behring Breivik as he lit the fuse on his homemade fertilizer bomb: Did he hesitate? Did he understand that he was saying goodbye to society forever?

I regularly feel like I'm standing on a diving board, daring myself to leap toward other people, trusting that it will be fine, that I'll be accepted. I'm terrified and have a knot in my stomach, but I let go. There are people who care for me.

I am the invisible child; I hope you see me. I laugh out loud, and I'm not as scared anymore. I am the Groke dancing happily on the beach, no longer spreading cold all around me.

In the evenings, my daughter and I read Moomin books in her bed. I lie right beside her as she nestles into the crook of my arm. The light from the lamp on the windowsill makes the bed feel like an illuminated boat floating on the dark water of an ice-cold sea. It's like we're sailing into a dark future together without knowing what's in store for us, if we'll make it, if we'll be alright.

Lately, she has been able to read increasingly long sentences, and one evening reads *Who Will Comfort Toffle?* to me. "Poor Miffle, always quick to scare, is even quicker to console, now Groke's no longer there," she reads. And it's then I suddenly realize what the book is about: Toffle feels comforted by *comforting someone else*, Miffle. No one is going to *give him* anything, but making himself strong for somebody else frees him from the anxiety and loneliness. I have always felt that it's me who needs to be comforted, and that only then will I feel whole again. But it's when I protect and comfort my child, or a neighbor or friend, that the ache in my stomach fades and disappears. I am not alone, and that doesn't just mean helping those around me, it means that when I need them to, they will also help me. Because we are all interwoven by eye contact and bodies, by stories and dreams. We are never alone. Our flock is humanity.

"That's enough for tonight," I finally say, closing the book. But then she wants to play blabb.

"Please, just for a bit!" she begs.

"Okay," I say, and she snuggles closer into the crook of my arm.

"What is blabb?" she asks, giving me hints and clues so I can guess what she's thinking. (Blabb can be anything.)

"Blabb is yellow and round and is on my desk," she says.

"Okay, it's the Easter egg we still haven't tidied away, even though Easter was months ago," I say.

"Yes. What's blabb?" she asks, moving on to the next round.

"We really have to go to sleep soon," I say, as though we're one person, not two.

"Just one more, Mom," she pleads. Her little game can keep us going for hours, she loves it.

"What's blabb? It's cool and nice, and it's the best thing ever!" she hints.

"Is it the cat?" I ask.

She shakes her head, smiles wryly; her eyes widen with the secret that's bursting to come out.

"Is it eating candy on Saturdays?" I guess again.

She smiles again, and looks at me full of joy:

"No! *No!* Don't you get it? Blabb is my *family!*"

Notes

Full notes for *So Lonely* can be found at this URL:
https://greystonebooks.com/products/so-lonely